Withdrawn

11/9/11
5
נמ
5
14 day
11/11

The Inquisition of Climate Science

The
INQUISITION
of CLIMATE
SCIENCE

James Lawrence Powell

Columbia University Press / New York

Columbia University Press
Publishers Since 1893
New York   Chichester, West Sussex
Copyright © 2011 James Lawrence Powell
All rights reserved

Library of Congress Cataloging-in-Publication Data
Powell, James Lawrence, 1936–
The inquisition of climate science / James Lawrence Powell.
p. cm.
Includes bibliographical references and index.
ISBN 978-0-231-15718-6 (cloth : alk. paper) — ISBN 978-0-231-52784-2 (ebook)
1. Global warming.   2. Climatic changes.   I. Title.
QC981.8.G56P69 2011
363.738'74—dc23
2011018611

∞

Columbia University Press books are printed on permanent and durable acid-free paper.
This book is printed on paper with recycled content.
Printed in the United States of America
c 10 9 8 7 6 5 4 3 2 1
References to Internet Web sites (URLs) were accurate at the time of writing. Neither the author nor Columbia University Press is responsible for URLs that may have expired or changed
since the manuscript was prepared.

On page ix the epigraph by Carl Sagan is from a speech he delivered in 1995; see http://seti
.sentry.net/archive/public/1999/5–99/00000351.htm. The epigraph by Richard Feynman is
from "Personal Observations on the Reliability of the Shuttle," *Appendix F, NASA Challenger
Report* (1986); see http://science.ksc.nasa.gov/shuttle/missions/51-l/docs/rogers-commission/
Appendix-F.txt.

To

James E. Hansen,

Michael E. Mann,

Benjamin D. Santer,

and the late Stephen H. Schneider

SCIENTISTS OF

COURAGE AND INTEGRITY

Contents

We have designed a civilization based on science and technology and at the same time have arranged things so that almost no one understands anything at all about science and technology. This is a clear prescription for disaster. We may for a while get away with this mix of ignorance and power but sooner or later it is bound to blow up in our face.

—CARL SAGAN

Nature cannot be fooled.

—RICHARD FEYNMAN

Preface

After a career as geology professor, college president, and museum director, I began to write science books for the general reader. Each book addresses an important question: Did a meteorite collide with Earth and kill the dinosaurs? How did the twentieth-century revolutions in geology—deep time, plate tectonics, and meteorite impact—come about and what took them so long? What process created the Grand Canyon? How did America come to build the great dams on the lower Colorado River and what is their future?

This book is about another question: why, when the scientific evidence for global warming is unequivocal, does only half the public accept that evidence? What has caused so many to doubt the conclusions of scientists, whom the public usually trusts, on such an important issue? In trying to answer that question, I have come to believe that in the denial of global warming, we are witnessing the most vicious, and so far most successful, attack on science in history. Never were the words of Thomas Jefferson more apt: "If a nation expects to be ignorant and free, in a state of civilization, it expects what never was and never will be."

What are my credentials? I am not a climate researcher. I like to think that may be an advantage, as I have no axe to grind, no position to defend. I do have a PhD in geochemistry from MIT. I have received research grants and written scientific articles and books. President Ronald Reagan and President George H. W. Bush each appointed me to the National Science Board, where I served for twelve years. That experience informed me about how science works at the level of national policy. Ultimately, of course, any book has to speak for itself.

The Inquisition of Climate Science

Introduction

Radio and television broadcasters accuse climate scientists of promoting a global warming hoax, recommending that they be "named and fired, drawn and quartered" (Rush Limbaugh); commit "hara-kiri" (Glenn Beck); and be "publicly flogged" (Mark Morano). The Viscount Monckton of Brenchley (Christopher Monckton) calls climate scientists evil and pronounces them as guilty of genocide as war criminal Radovan Karadžic.

Gerald Warner, a columnist for the UK *Telegraph*, writes that

> the status of the white-coated prima donnas and narcissists has never been lower. . . . After a period of priest-like authority, the pointy-heads in lab coats have reassumed the role of mad cranks. . . . The public is no longer in awe of scientists. Like squabbling evangelical churches in the 19th century, they can form as many schismatic sects as they like, nobody is listening to them any more. Unquestioned authority derived from a white coat and a doctorate is as dead as the Druids.[1]

Politicians are not shy about joining in. On June 26, 2009, the U.S. House of Representatives took up the first major climate bill to have a chance of passing the Congress: the American Clean Energy and Security Act (ACES), known as the Waxman-Markey bill for its two sponsors. When it came his turn to speak, Rep. Paul Broun (R-GA), who trained as a medical doctor, said, "Scientists all over this world say that the idea of human-induced global climate change is one of the greatest hoaxes perpetrated out of the scientific community. It is a hoax. There is no scientific consensus."[2] After his remarks, Broun's Republican colleagues gave him a hearty round of applause.

On February 8, 2010, former vice presidential candidate and onetime Alaska governor Sarah Palin labeled global warming a "bunch of snake oil science." Sen. James Inhofe (R-OK) denounces global warming as "the greatest hoax ever perpetrated on the American people," recommending that to get "the real story" people should read Michael Crichton's novel, *State of Fear*.

Also in February 2010, several months after climate-related e-mails stolen from the University of East Anglia surfaced (see chapter 14), Senator Inhofe accused climate scientists of

Obstructing release of damaging data and information; manipulating data and knowingly using flawed climate models to reach preconceived conclusions; colluding to pressure journal editors who published work questioning the climate science "consensus"; and assuming activist roles to influence the political process.[3]

Inhofe's statement lists seven federal statutes that climate scientists named in the stolen e-mails may have violated, including the False Claims Act and Obstruction of Justice and Interference with Federal Proceedings. The Inhofe statement concludes: "The CRU [Climate Research Unit of the university] documents and emails reveal, among other things, unethical and potentially illegal behavior by some of the world's preeminent climate scientists." The report identifies seventeen, a who's who of climate science, as "key players."

Science has been under attack before, often because of a perceived conflict with religion. The man whom physicist Stephen Hawking says is more responsible for the birth of modern science than anyone, Galileo, was the victim of the Roman Inquisition. During the sixteenth century, the Catholic Church staged trials of those it accused of heresy, sorcery, immorality, blasphemy, Judaizing, and witchcraft. For espousing the view that the Sun and not Earth lies at the center of the solar system, the Inquisition found Galileo guilty of heresy, forced him to recant, and sentenced him to prison, later commuting the sentence to lifelong house arrest. Galileo had friends in high places, otherwise his fate might have been the same as that of the many others whom the Inquisition burned at the stake.

Global warming deniers submit climate scientists to a modern inquisition conducted not in a courtroom but on the front pages of newspapers, on right-wing radio and television, on the blogosphere, and on denier websites. Just as the Roman Inquisition rejected outright the evidence that Copernicus and Galileo had assembled to show that Earth moves around the Sun, the modern inquisitors deny the overwhelming evidence of global warming. Galileo's trial ended; the modern inquisition of climate science shows no sign of abating; indeed, as the evidence for global warming mounts, the attacks on scientists grow louder.

At least the Roman Inquisition had an alternative theory of the solar system: Ptolemy's earth-centered astronomy from the second century C.E. Even today's creationists and disciples of intelligent design have their Bible to fall back on. The modern inquisitors have not even that much: they have no alternative theory to explain the observed facts of global warming. As Peter Gleick, president of the Pacific Institute put it, "Those who deny that humans are causing unprecedented climate change have never, ever produced an alternative scientific argument that comes close to explaining the evidence we see around the world that the climate is changing."[4]

How long will we let a campaign of denial delay action to limit the damage from global warming? How long will thoughtful people stand on the sidelines while deniers vilify scientists and attempt to neutralize an entire field of science? How long will the media continue to present "both sides" when the science on one side is overwhelming and the other side has neither facts nor theory?

These questions loom larger because climate science is not the only discipline under attack. Creationists and followers of intelligent design continue to deny evolution. In *Doubt Is Their Product*, David Michaels shows how corporations have fostered and supported anti-science campaigns to deny the dangers of asbestos, chromium, lead, plastics, aspirin use by children, and more.[5] Seth Kalichman, in his important book *Denying Aids*, chronicles the AIDS denialists, whose tactics eerily mimic those of the global warming deniers.[6] In *Autism's False Prophets*, Paul Offit reviews the history of what we might call medicine-denial, exposing the false claim that vaccinations cause autism. History deniers claim that the Holocaust and the Moon landings never happened. Conspiracy theorists argue that the United States, or Israel, destroyed the World Trade Towers; and that yet-to-be-uncovered plots led to the murders of John F. Kennedy, Robert F. Kennedy, Martin Luther King, Marilyn

Monroe, Princess Diana, and Vince Foster. Science, history, and medicine are all under attack. Indeed, science denial is so widespread that reason itself is threatened.

Not Skeptics, Deniers

There is skepticism; then there is denial. The honest skeptic plays an essential role in science. Some four centuries ago, Sir Francis Bacon captured the proper attitude: "If a man will begin with certainties, he shall end in doubts; but if he will be content to begin with doubts, he shall end in certainties[7]. In 1910, T. H. Huxley wrote: "[For the] improver of natural knowledge skepticism is the highest of duties, blind faith the one unpardonable sin."[8]

Scientists need to be cautious, reserving judgment until they fully corroborate new claims. Extraordinary claims, as Carl Sagan said, require extraordinary evidence. But those who still refuse to accept the evidence for global warming are not skeptics. When evidence becomes strong enough, an honest skeptic is honor bound to accept it. But as the evidence of global warming has accumulated relentlessly over the last twenty years, one group has remained obstinate, not only refusing to accept the new evidence, but becoming ever more outspoken against it. Global warming deniers do not merely question the evidence and ask for more and better research, much less present any of their own. Rather they denounce climate science and those who practice it, ridiculing them and questioning their ethics and honesty. Congressional committees call scientists to testify and subject them to a Kafkaesque interrogation.

To call such people skeptics is to sully a term of honor, allowing the deniers to cloak themselves in the mantle of science even as they deny critical parts of climate science. Those who abjure global warming are not skeptics; they are *deniers*. To call them skeptics is to debase language as much as to call the Ku Klux Klan "prejudiced," Holocaust deniers "biased," or Flat-Earthers "mistaken." There is honest and honorable skepticism; then there is irrational, self-serving denial.[9]

Note: The Intergovernmental Panel on Climate Change reported in 2007 that global warming is "unequivocal" and that it is "very likely," defined as greater than 90 percent probable, that humans are the cause. Three years later, it is clear that the IPCC underestimated the extent and the threat of global warming. In this book, instead of writing "anthropogenic global warming," or "human-caused global warming," I will simply write "global warming."

Science and Potemkin Science

Anyone following the controversy over global warming is bound to come away with the impression of two parallel but separate universes: the scientists and the deniers. Two conferences, one held in December 2008 and the other in March 2009, reinforced that impression. The two meetings were outwardly identical, as speakers illustrated their remarks with charts and tables and took questions and comments from their audience. But there the resemblance ended, for the two sets of speakers began with different missions and ended with opposite conclusions.

The scientists who presented their research at the December 2008 meeting of the American Geophysical Union (AGU), held in San Francisco's Moscone Center, evidently regarded global warming as an observational fact.[1] Dr. James Hansen, of NASA's Goddard Institute for Space Studies at Columbia University, gave an invited lecture titled, "Climate Threat to the Planet: Implications for Energy Policy and Intergenerational Justice." Sixteen thousand AGU members had come to the Bay City for the meeting, and Hansen's audience filled the conference room. He had become the most authoritative and outspoken scientist on global warming, warning with increasing urgency that rising temperatures threaten the future of humanity. Twenty years earlier, Hansen had been one of the first to sound the alarm, testifying on a sweltering June day to members of the Senate Energy and Natural Resources Committee that he was 99 percent certain that global warming had begun. "It is time to stop waffling so much and say that the evidence is pretty strong that the greenhouse effect is here," Hansen cautioned.[2] Now, two decades later, for Hansen and his AGU colleagues, the evidence for global warming had grown from "pretty strong" to virtually certain.

Hansen's science and his forthrightness had earned him the respect of his peers, election to the National Academy of Sciences, the AGU Award for "Scientific Freedom and Responsibility," as well as medals from the World Wildlife Fund and the American Meteorological Association, the latter for "outstanding contributions to climate modeling, understanding climate change forcings and sensitivity, and for clear communication of climate science in the public arena." EarthSky Communications and a panel of 600 scientist-advisers named Hansen as their "Scientist Communicator of the Year," praising him as an "outspoken authority on climate change" who had "best communicated with the public about vital science issues or concepts during 2008."[3] In June 2010, Hansen won the Sophie Prize, set up in 1997 by Norwegian Jostein Gaarder, the author of the 1991 best-selling novel and teenagers' guide to philosophy, *Sophie's World*, for Hansen's "key role for the development of our understanding of human-induced climate change."

But the more awards Hansen accumulated and the more his climate forecasts turned out to be correct, the more the deniers disparaged him. They had no choice, for if Hansen is right, the deniers are wrong. One can even go to Facebook.com and sign a petition asking NASA to fire Hansen.

In his AGU talk, the NASA scientist showed a series of slides summarizing the state of climate science at the end of 2008. Some images came from his own research, but most were the work of other scientists. The overall impression was of global warming advancing more rapidly than scientists had suspected only a few years earlier. One slide was unusual for a scientific talk: a photograph of Hansen's newest grandchild, Jake, whom he said "will live in the greenhouse world that we choose to create."

Another speaker at the 2008 AGU meeting, Wallace Broecker of Columbia University, is also a pioneer of climate science. In 1975, Broecker published a paper in *Science* titled, "Climatic Change: Are We on the Brink of a Pronounced Global Warming?"[4] Broecker had made many pathbreaking discoveries, becoming best known for his idea that ocean currents not only operate on the surface, but at depth as a kind of conveyor belt that transports salty, dense seawater around the globe.[5] In the North Atlantic, winds push surface currents from the equator toward the poles, where they cool and sink, the deep water flowing in the opposite direction all the way to Antarctica. The oceanic conveyor belt helps control global climate, leading Broecker to fear that as the planet warms, melting freshwater ice might dilute the salty current and shut down the conveyor, changing the climate of the North Atlantic in unpredictable ways.

But in his AGU talk, titled "Shifting Rainfall: A Paleo Perspective," like an honest scientific skeptic, Broecker found his own theory wanting and rejected it: "Twenty years ago my concern regarding the impacts of the ongoing CO_2 buildup were centered on the ocean's conveyor circulation. Would the predicted increase in rainfall and runoff lead to a sudden shutdown? In the meantime, model simulations have made clear that this is highly unlikely."[6]

In the AGU conference session titled "Global Environmental Change," scientists presented more than 100 papers describing new research results. Not one contradicted the greenhouse theory of global warming. To say that the AGU scientists shared a consensus on global warming would be an understatement. Unanimity would be more accurate.

Never a Crisis

In March 2009, 800 "scientists, economists, legislators, policy activists, and media representatives" gathered in New York for a three-day conference sponsored by the Heartland Institute.[7] According to the institute, "The presenters at this year's conference are the elite in the world among climate scientists." If that were true, then many of the speakers would be active in original climate research, publishing their results in peer-reviewed scientific journals. But almost none of the speakers met that description. Many did not even have degrees in science. Still, they included a distinguished MIT meteorologist, the last man to set foot on the Moon, and the president of the Czech Republic.

The speakers scorned global warming not only as false, but a deliberate hoax designed, as the former astronaut put it, to effect "an enormous transfer of wealth from the people to the government."[8] The title of the conference presaged its conclusions: "Global Warming: Was it Ever a Crisis?"

The mission of the Heartland Institute is "to discover, develop, and promote free-market solutions to social and economic problems."[9] As I will show later, the Heartland Institute is an anti-regulation organization historically funded by Big Tobacco and Big Oil, with an avowed mission and history of promoting the interests of corporations.

Not only did the conclusions of the speakers at the two conferences differ, so did their methods. Having few if any research results of their own to report, in contrast to the AGU speakers, presenters at the Heartland Institute Conference attacked the research findings of mainstream scientists, looking for any discrepancy or inconsistency. The speakers said that global warming is natural, not man-made; that atmospheric carbon dioxide does not cause global warming; that global warming will not harm humans or coral reefs or cause extreme weather events. Computerized climate models do not work, the presenters claimed, and the alleged scientific consensus on global warming is nothing more than an "urban myth." And that list comes from only the first morning of the conference. After the meeting, the Heartland Institute answered its own question of whether global warming had ever been a crisis with "a resounding 'no.'"[10]

A Political Movement

One of the keynote speakers at the Heartland conference, Richard Lindzen, the Alfred P. Sloan Professor of Meteorology at MIT, does have outstanding scientific credentials, though most of his research has not dealt directly with global warming. Lindzen titled his talk: "Climate Alarm: What We Are Up Against, and What to Do." He opened with these words: "Global warming alarm has always been a political movement," thus preemptively denouncing those who disagree with him as motivated by politics, not science, a ubiquitous denier theme.[11]

Speakers at the Heartland Institute Conference indulged in personal attacks, especially on Hansen and former Vice President Al Gore, whose book and film, *An Inconvenient Truth*, had made him their bogeyman. One might have hoped that Lindzen, a distinguished professor, member of the National Academy of Sciences, and AGU medal winner himself, would have risen with dignity above petty personal attacks, but no, he got down in the mud, accusing scientists who espoused global warming—even his own MIT colleagues—of dishonesty and worse: of selling out their scientific integrity for money. Lindzen named names:

> Most of the atmospheric scientists who I respect do endorse global warming [but] the science that they do that I respect is not about global warming. Endorsing global warming just makes their lives easier. My colleague, Kerry Emanuel, received relatively little recognition until he suggested that hurricanes might become stronger in a warmer world. He then was inundated with professional recognition. Another colleague, Carl Wunsch . . . [has] politics [that] are clearly liberal. Wally Broecker['s] work clearly shows that sudden climate change occurs without anthropogenic influence. However, he staunchly beats the drums for alarm and is richly rewarded for doing so.[12]

In his speech, Lindzen neither presented research results of his own nor did he explain why for twenty years he has been adamant that global warming is false. He did, however, say that "Nature is, as any reasonable person might suppose, dominated by stabilizing negative feedbacks rather than destabilizing positive feedbacks." Negative feedbacks move the system back toward stability. Positive feedbacks amplify the output of a system—and make global warming worse. Though half a century of research, including attempts by Lindzen himself, has failed to find significant negative feedbacks, Lindzen continues to claim that they exist in sufficient strength to render global warming harmless and allow business-as-usual to continue.

A Cheap Tuxedo

Another major difference between the conferences of scientists and those of deniers lies in the aftermath. When their meetings end, scientists rush back to their labs to continue their research and follow leads picked up at the meeting. Having no labs to go to, deniers conduct a public relations stunt by issuing a "declaration," a cleverly worded statement that dresses up their denial in fancy duds. To paraphrase Kris Krishtalka, a denier declaration is nothing more than anti-science in a cheap tuxedo.

After a 1992 meeting in Germany just prior to the Rio Earth Summit, deniers put out the Heidelberg Appeal, saying, "We are . . . worried . . . at the emergence of an irrational ideology which is opposed to scientific and industrial progress and impedes economic and social development." Three years later came the Leipzig Declaration, which proclaimed, "There does not exist today a general scientific consensus about the importance of greenhouse warming from rising levels of carbon dioxide. On the contrary, most scientists now accept the fact that actual observations from earth satellites show no climate warming whatsoever." Richard Lindzen signed both declarations.[13]

After its conference in March 2008, the Heartland Institute upheld the tradition by issuing a "Manhattan Declaration on Climate Change." Among its conclusions were that

> Carbon dioxide is not a pollutant but rather a necessity for all life . . . assertions of a supposed "consensus" among climate experts are false . . . warmer weather is generally less harmful to life on Earth than colder . . . there is no convincing evidence that CO_2 emissions from modern industrial activity has in the past, is now, or will in the future cause catastrophic climate change. . . . We recommend that world leaders reject the views expressed by the United Nations Intergovernmental Panel on Climate Change as well as popular, but misguided works such as "An Inconvenient Truth" [and] that all taxes, regulations, and other interventions intended to reduce emissions of CO2 be abandoned forthwith.[14]

As we will learn, scientists have long disproven each of the alleged statements of fact in that declaration. Moreover, since the Intergovernmental Panel on Climate Change (IPCC) and Al Gore jointly won the 2007 Nobel Peace Prize, there was little chance that world leaders would reject their conclusions.

These declarations are most useful in revealing how the deniers operate. They

- Engage in publicity stunts designed to gain media attention and that promulgate disinformation.
- Repeat claims long after scientists have shown them to be false.
- Make assertions without presenting any evidence to back them up. Had a speaker at the AGU meeting said that carbon dioxide does not cause global warming, the audience would have demanded to see the evidence.

- Have no scientific findings that falsify global warming.
- Have opposed global warming for twenty years. True, back then, many scientists were also skeptical, but as the evidence mounted, they changed their minds. Deniers do not change their minds, a sure sign that they base their denial not on science, but on ideology. To paraphrase Lindzen, global warming denial has always been about politics, not science.

The 2008 statement is also useful in revealing the true motivation of organizations like the Heartland Institute: to prevent the use of "taxes, regulations, and other interventions" to reduce CO_2 emissions. The Heartland Institute may even have no particular beef with global warming; its objection is to taxes and government regulations that might crimp American corporations. This explains why the institute and like-minded groups consistently wind up on the side of corporations and opposing science.

Over the fall of 2008 and the ensuing winter that preceded the March 2009 Heartland Institute Conference, a number of new findings showed global warming moving faster than scientists had expected even a year or two earlier. Speakers at the conference had access to this new evidence, but it had no effect on their presentations. Everything they said at the 2009 Heartland conference they could have said at the 2008 conference—and many did—and said again at the 2010 conference.

Consider this list of new findings:

- *Nature* and *Science*: Sea level "most likely" to rise 0.8 to 2.0 meters by 2100 (Fall 2008).
- U.S. Geological Survey: Sea-level rise in 2100 will likely "substantially exceed" IPCC projections (December 2008).
- Britain's Hadley Center: "Catastrophic 5–7°C (9–12.6°F) warming by 2100 on current emissions path" (December 2008).
- Australian newspaper reports: Worst drought in Australia's history. On January 29, 2009, South Australia had its hottest nighttime temperature on record. Wildfires soon devastated the region and over 200 died, some incinerated in their cars. While global warming may not have caused the fires, they certainly demonstrate the risk.
- MIT: Projection of global warming by 2100 doubles to 5.1°C (9.2°F)" (February 2009)

- American Association for the Advancement of Science: Climate change is coming much harder, much faster than predicted (February 2009).
- National Oceanographic and Atmospheric Administration (NOAA): Climate change "largely irreversible for 1,000 years," with persistent and pervasive droughts around the globe (February 2009).
- "Worst-case IPCC scenario trajectories (or even worse) are being realized." "Key Message" from the Copenhagen 2009 International Scientific Congress on Climate Change (March 2009).
- NOAA: January and February 2009 warmer than in any year on record (March 2009).
- Environmental Protection Agency (EPA): Global warming a public danger (March 2009).

A ubiquitous theme, and the party line of the 2009 Heartland Institute Conference, is that the IPCC deliberately exaggerated the evidence for dangerous global warming. But as these new findings show, global warming was moving faster than the IPCC had estimated in its 2007 Assessment Report.

Better Than Scientists

Science seeks the truth about Nature. In the process, scientists follow false trails and have to backtrack. They make mistakes; groupthink can blinker them from the truth. But the scientific method eventually exposes the errors. Even then, some cling to the old ways, taking their outmoded beliefs with them to the grave.

Most professions can be no better than their individual practitioners, but Science is far better than scientists. It is the best system we have for getting beyond human frailty and folly to the truth. Those who trample science are always in time the ones to suffer. Who stands guilty in the courtroom of history: Galileo, or his inquisitors?

Adventures in Denierland

To meet a global warming denier and begin to understand their organizations and tactics, let us look in depth at the presentation at the 2009 Heartland Institute Conference by Terry Dunleavy, MBE, JP, titled, "'Consensus' in Climate Science: An Unsubstantiated Urban Myth."[1] (MBE stands for "Member of the British Empire," the most junior of the British Orders of Chivalry; JP for Justice of the Peace.) In probing Dunleavy's claims, we will wander in and out of the Internet to visit other deniers and their websites.

Dunleavy, a congenial, 80-year-old New Zealand wine grower and self-professed nonscientist, serves as executive vice chairman of the International Climate Science Coalition (ICSC), whose website describes it as "an international association of scientists, economists and energy and policy experts working to promote better public understanding of climate change science and policy worldwide."[2] Fair enough; if there is one thing clear from this book, it is that science needs better public understanding. But the website's next sentence reveals that the ICSC is dedicated to "providing a highly credible alternative to the UN's Intergovernmental Panel on Climate Change (IPCC) thereby fostering a more rational, open discussion about climate issues."[3] Were there any doubt as to its mission, the ICSC's "core principles" remove it, giving us an advance look at the main denier arguments against global warming. (The phrases in parentheses below are my summaries of the claims of the ICSC.)

1. "Global climate is always changing in accordance with natural causes and recent changes are not unusual." (Global warming is natural.)

2. "Science is rapidly evolving away from the view that humanity's emissions of carbon dioxide and other 'greenhouse gases' are a cause of dangerous climate change." (Scientists are changing their minds about the cause and the dangers of global warming.)

3. "Climate models used by the IPCC fail to reproduce known past climates without manipulation and therefore lack the scientific integrity needed for use in climate prediction and related policy decision-making." (Climate models don't work.)

4. "The UN IPCC Summary for Policymakers and the assertions of IPCC executives too often seriously misrepresent the conclusions of their own scientific reports." (The IPCC cheats.)

5. "Claims that 'consensus' exists among climate experts regarding the causes of the modest warming of the past century are contradicted by thousands of independent scientists." (There's no consensus.)

6. "Carbon dioxide is not a pollutant—it is a necessary reactant in plant photosynthesis and so is essential for life on Earth." (Carbon dioxide is good for us.)

7. "Research that identifies the Sun as the principal driver of global climate must be taken more seriously." (It's the Sun.)

8. "Global cooling has presented serious problems for human society and the environment throughout history while global warming has generally been highly beneficial." (Warmer is better.)

9. "It is not possible to reliably predict how climate will change in the future, beyond the certainty that multi-decadal warming and cooling trends, and abrupt changes, will all continue, underscoring a need for effective adaptation." (Predicting the future is futile. Instead of trying to prevent global warming, society should learn to adapt.)

10. "Since science and observation have failed to substantiate the human-caused climate change hypothesis, it is premature to damage national economies with 'carbon' taxes, emissions trading or other schemes to control 'greenhouse gas' emissions." (Wait for certainty; meanwhile, don't do anything because it would be too expensive.)

Contrast the mission and principles of the ICSC with those of the AGU, which "fosters excellent Earth and space science research, to the benefit of humanity." The AGU's first two "values" are "the scientific method" and "the generation and dissemination of scientific knowledge." The AGU and its members seek answers; denier organizations like the ICSC know the answers and seek only confirmation that they are right. One group of minds is open; the other closed.

Urban Myth?

Dunleavy begins his attack by listing three groups that help to establish the "urban myth" of scientific consensus:

- The IPCC, whose reports represent the work of 2,500 scientists
- Science academies and organizations
- Many individual scientists

Dunleavy turns first to the IPCC. Obviously, a much smaller group than the organization's 2,500 scientists must draft its important Summary for Policymakers. After that, government officials from the member nations edit the scientists' language. Both approaches give the deniers an opening that they use to try to discredit the IPCC.

Dunleavy bases his presentation not on his own work, but on a study of the IPCC's Fourth Assessment Report by "climate data analyst" John McLean, a member of the policy board of ICSC and of another denier organization, the New Zealand Climate Science Coalition. The mission of the NZCSC is "To represent accurately, and without prejudice, facts regarding climate change; to provide considered opinion on matters related to both natural and human-caused climate effects."[4] Again, who could object? But read a bit further and you find that the organization's members are "concerned at the misleading information being disseminated about climate change and so-called anthropogenic (man-made) global warming."

McLean has studied the IPCC, its organization, and its methodology with the punctiliousness of a Talmudic scholar, attempting to show that many of the 2,500 IPCC scientists do not endorse its overall conclusions. Following suit, Dunleavy criticizes the IPCC process because "Only 51 authors worked on the draft version" of the Summary for Policymakers (SPM), written "at a plenary session primarily of government officials and representatives of environmental and industry organizations." Fifty-one would seem a more than adequate number of authors, if not far too many, so Dunleavy moves on to argue that the SPM actually represents the views of government bureaucrats, who twist the scientists' language to make it more extreme on global warming. At least four things are wrong with this claim.

First, the larger a group, the more conservative its statements. Extreme positions do not survive a process of drafting by 51 scientists from different disciplines and countries, with government bureaucrats having final oversight. The IPCC is by its nature a conservative organization prone to understatement. This is not only true in theory, it is obvious when we compare the current evidence for global warming with the statements of the IPCC in its 2007 report.

Second, in May 2001 the White House asked the National Academy of Sciences to identify "areas in the science of climate change where there are the greatest certainties and uncertainties." The request also sought the views of the Academy "on whether there are any substantive differences between the IPCC reports and the IPCC summaries."[5] The Academy committee "solicited written responses from U.S. coordinating lead authors and lead authors of IPCC chapters, reviewed the . . . draft report and summaries, and interviewed . . . [a] coordinating lead author for the IPCC [Working Group] Technical Summary." From that analysis, the Academy concluded "that no changes were made without the consent of the convening lead authors and that most changes that did occur lacked significant impact." The panel found that, "The IPCC's conclusion that most of the observed warming of the last 50 years is likely to have been due to the increase in greenhouse gas concentrations accurately reflects the current thinking of the scientific community on this issue." Eight years before Dunleavy's presentation, an Academy panel had already falsified his claims.

Third, consonant with scientific skepticism, scientists are not shy about questioning authority. If large numbers of scientists believed that the IPCC process is corrupt, one would only have to pick up the latest newspaper to find it out. In January 2006, the press leapt on James Hansen's claim that NASA officials were attempting to muzzle him, as they did over similar allegations from scientists at NOAA.[6]

Fourth, the claim that government bureaucrats squelch IPCC scientists in order to promulgate the bureaucrats' extremist views on global warming is untrue. Does anyone believe that American officials during the presidency of George W. Bush tried to *strengthen* scientists' statements on global warming? Governments created the IPCC to prevent scientists from making public statements that would lead to action before governments were ready. Were there any doubt that government representatives weaken, not strengthen, the scientists' conclusions, consider this evidence from the IPCC's Fourth Assessment. After the scientists had completed their draft, diplomats and bureaucrats took over and began to water down the conclusions. A word-for-word comparison between the scientists' draft report and the subsequent official IPCC version goes on for pages, revealing scores of instances where the bureaucrats weakened the scientists' conclusions.[7] Not once did government officials strengthen the wording. Here is one example from many:

> *Scientists:* "Impacts are very likely to increase due to increased frequencies and intensities of extreme weather events [high confidence]."
> *Final Summary for Policymakers:* "Impacts due to altered frequencies and intensities of extreme weather, climate, and sea level events are very likely to change."

Dunleavy next moves to the "Independent Summary for Policymakers" or ISPM, a denier critique, prepared by nine "climate experts," of the IPCC's Fourth Assessment Report. The coordinator of the Independent Summary for Policymakers is Ross McKitrick, PhD and Associate Professor of Economics at the University of Guelph in Ontario and Senior Fellow, Fraser Institute, Vancouver, British Columbia. The motto of the Fraser Institute is "a free and prosperous world through choice, markets and responsibility." Of global warming, the institute says, "Scientific evidence about the extent and cause of climate change continues to advance, but significant uncertainties remain. In attempting to pressure policy decisions, some activist groups risk exaggerating the certainty and the damages of human impacts on future climate change."

McKitrick became one of the best-known global warming deniers, even drawing an invitation to testify before Senator Inhofe's committee, for his criticism of an icon of global warming, the so-called "hockey stick" reconstruction of global temperatures over the past 1,000 years by climate scientists Michael Mann, Raymond Bradley, and Malcolm Hughes, which I discuss in chapter 12.[8]

The Independent Summary for Policymakers (ISPM) is the deniers' attempt to rebut and discredit the IPCC's Summary for Policymakers.[9] The website of the ISPM identifies its nine authorities (whose individual accomplishments are noted in the following list), allowing the scientists at RealClimate.org to review the publication record of each.[10] (RealClimate.org is a website and blog produced and maintained by a group of practicing climate scientists.)

1. Published a bulletin on a meteorology meeting in 1994.
2. Published an anti–global warming review in 2003 but has done no original research.
3. Published "Don't be Gored into Going Along" in *Power Engineer* in 2006. No other publication since 1973.
4. Publishes on chaos and predictability, not on climate research.
5. Publishes on paleoclimates, in particular on dating techniques.
6. Published papers on remote sensing.
7. Published papers on climate records from ice cores.
8. Publishes on tsunamis and storm surges.
9. No publication in the last decade.

RealClimate.org continues with a line-by-line critique of the Independent Summary for Policymakers, finding an error in nearly every paragraph. The ISPM description of greenhouse warming is "nonsense"; the report "misrepresents the recent National Research Council report" which affirms that warming over the last 1,000 years is anomalous; flings mud at the climate models in hope that some accusations stick; falsely accuses the IPCC of giving "limited consideration" to aerosols; incorrectly suggests that snow cover has increased; and so on. Realclimate. org concludes, "There are so many bizarre statements in the [ISPM] that spotting them could serve as a good final exam in an elementary course of climate change."

Cave Junction

Dunleavy quickly passes over the role of science societies in demonstrating consensus, as well he must, for who would attack directly the AGU, the U.S. National Academy of Sciences, and the scores of other national and international scientific organizations that have issued statements and declarations accepting global warming? To attack them would be to attack science itself. (The Appendix in this book lists the thirty-three national science academies and the nearly seventy international science organizations that have issued statements accepting global warming.)

Dismissing these societies in a few words, Dunleavy speeds on to his third group: individual scientists. The deniers attempt to show that large numbers of scientists do not support the alleged consensus, not by citing peer-reviewed articles that question global warming, but by exhibiting signatures on petitions. The most infamous comes from the Oregon Institute of Science and Medicine, signed as of May 13, 2010, by 31,486 "scientists, engineers and other technically trained professionals." According to the OISM, the signatories include Edward Teller, "Father of the H-Bomb," who asked to have more petition cards sent to him so that he could distribute them. Who could resist following the Internet trail to learn more about the Oregon Institute of Science and Medicine?

SourceWatch.org tells us that the OISM resides on a farm seven miles from the town of Cave Junction, Oregon (population 1,126), south of the famous Rogue River in one of the more remote stretches of the American West. (SourceWatch is a project of the Center for Media and Democracy, "an independent, non-profit, non-partisan consumer and citizen watchdog group.") Heading OISM is Dr. Arthur B. Robinson, according to SourceWatch "an eccentric scientist who has a long history of controversial entanglements with figures on the fringe of accepted research." In addition to its denialist activities, the OISM provides homeschooling tips for parents who are worried about socialism in the public schools and publishes books on how to survive a nuclear war. As SourceWatch notes, "Cave Junction is the sort of out-of-the-way location you might seek out if you were hoping to survive a nuclear war, but it is not known as a center for scientific and medical research." The institute's website lists eight faculty members, including Arthur Robinson and his two sons, Noah and Zachary. A fuzzy photograph on the institute's home page shows a large warehouse-like building bearing a sign on which one can dimly perceive the name Oregon Institute of Science and Medicine.[11] Dr. Robinson and two other institute faculty members, Dr. Martin Kamen and Dr. R. Bruce Merrifield, who won the 1984 Nobel Prize in Chemistry, pose in front of the building.

The OISM would likely be as obscure as its location were it not for its role in circulating the aforementioned petition, mailed in 1998 to thousands of scientists under the imprimatur of Dr. Frederick Seitz, former president of the National Academy of Sciences and President Emeritus of Rockefeller University. The petition statement begins by urging the United States government to reject the Kyoto Protocol. The second paragraph states, "There is no convincing scientific evidence that human release of carbon dioxide, methane, or other greenhouse gases is causing or will, in the foreseeable future, cause catastrophic heating of the Earth's atmosphere and disruption of the Earth's climate. Moreover, there is substantial scientific evidence that increases in atmospheric carbon dioxide produce many beneficial effects upon the natural plant and animal environments of the earth."[12]

According to SourceWatch, after some 15,000 had signed the petition, Sen. Chuck Hagel (R-NE) noted the "extraordinary response" and credited the petition with strengthening his opposition to an international global warming treaty. Hagel went on to say that, "Nearly all of these 15,000 scientists have technical training suitable for evaluating climate research data." Publications as diverse as *Newsday*, the *Los Angeles Times*, the *Washington Post*, the *Austin-American Statesman*, the *Denver Post*, and the *Wyoming Tribune-Eagle* cited the petition as a valid indicator of scientific opinion. On September 17, 2008, Bob Lutz, vice chairman of General Motors, said on cable TV's *Colbert Report*, "In the opinion of about 32,000 of the world's leading scientists [global warming is not real.]"[13]

Dr. Arthur Robinson has a distinguished background in chemistry, having been an associate of twice Nobel Prize winner Linus Pauling at the University of California at San Diego. Together, they established the Linus Pauling Institute of Science and Medicine to explore Pauling's idea that high doses of vitamin C prevent colds, mental illness, cancer, and many other diseases. Robinson's research showed that, to the contrary, too much vitamin C is harmful. Pauling called Robinson's work "amateurish." Robinson left Pauling's lab and settled in Cave Junction, where in 1980 he founded the OISM.

Robinson himself signed the "Scientific Dissent from Darwinism" petition of the Discovery Institute, the most prominent organization to deny evolution and promote intelligent design. The signers "are skeptical of claims for the ability of random mutation and natural selection to account for the complexity of life."

The letter from Frederick Seitz bore all the earmarks of a paper published in the peer-reviewed *Proceedings of the National Academy of Sciences*: same font, same format. The authors were Robinson, 22-year-old son Zachary, and two scientists, Sallie Baliunas and Willie Soon, well-known deniers who have worked for the George C. Marshall Institute and a number of other anti-science organizations. The paper, titled "Environmental Effects of Increased Atmospheric Carbon Dioxide," turned out not to be a reprint from *PNAS* or from any peer-reviewed journal: Robinson had typeset it on his home computer.

In spite of its checkered provenance, the OISM petition might have served a useful purpose by identifying scientists who were outside the mainstream on global warming. However, the way the OISM conducted the poll made that impossible to ascertain. Arthur Robinson admitted that only 2,100 signers had identified themselves as physicists, geophysicists, climatologists, or meteorologists, making it impossible to say how many came from the core disciplines of climate science, whose rejection of the consensus would be the most meaningful. The names are listed on the OISM website, but no institutional connections or home addresses are given—not even the city of residence. Added without incident were best-selling novelist John Grisham, several members of the original cast of the TV series *M*A*S*H*, and one Geraldine Halliwell, aka Ginger Spice (of the Spice Girls rock group), whose specialty was given as biology; but as Robinson acknowledged in an interview, "When we're getting thousands of signatures there's no way of filtering out a fake."[14]

Anyone can sign the petition by downloading a form from the OISM website and sending it in to the institute. The current petition does request address and degree.

Dunleavy's presentation is typical in that it reports no new science, nor does it discuss past scientific findings. Instead, Dunleavy focuses on the procedures of the IPCC and the number of signers of an online petition, which he uses to deny a scientific consensus on global warming. Dunleavy tries to prove a negative; let us examine the positive evidence for consensus.

The Evidence for Consensus

To understand the case for consensus, we need to know how science operates and how scientists acquire knowledge. They begin by making observations and developing hypotheses to explain them. Next they devise experiments or measurements to test the hypotheses. If enough tests corroborate a hypothesis, scientists may elevate it to the status of a theory. Contrary to popular usage, in science *theory* is a term of honor used to describe a concept for which there is considerable evidence, though not proof, and which explains enough of the observations to merit further testing and refinement. (One reason scientists never finally prove a theory is that new evidence could always appear and show that the theory is false.)

Spam Filter

Scientists practice their profession by doing research, which may include fieldwork, laboratory experiments, computer modeling, and so forth. They write up their methods and results and submit them to a scientific journal. The editor sends the paper out for review to several experts in the field—the "peers" of the author. The reviewers examine the paper to see whether the author has taken into account the relevant research in the subject, whether the research methods appear to be reliable and replicable, and whether the conclusions are reasonable. Reviewers do not try to replicate the research themselves, but rather to ensure that there is enough information so that experts in the field could replicate it. The reviewers can recommend that the journal publish the article as submitted, that it needs more research or a better write-up, or that the editor reject the article. Even the best articles are apt to have gone through at least one cycle of rewriting and resubmission.

Reviewers do not have to agree with every statement in an article or with its conclusion, but they must think it worthy of publication so that others may learn of the work and allow it to inform their own research. The process means that articles in peer-reviewed journals have no obvious defects and represent a contribution.

Some journals do not use peer-review, depending instead on the opinion of an editor or set of editors. One famous incident occurred in 1937, when the editor of the British journal *Nature* turned down a paper on the metabolic cycle of citric acid. Another journal published the paper; its author, Hans Krebs, went on to share the 1953 Nobel Prize in Medicine for his discovery of the "Krebs Cycle."

The experts who review manuscripts have no way of detecting outright fraud and some is bound to slip through. Especially in recent years, there have been a number of cases where authors turned out to have faked or plagiarized results, or simply made mistakes. The same is true of every field, one supposes, yet there is something different about science. The more important a result, the more other scientists will try to replicate it. A false claim is eventually, and sometimes rapidly, exposed.

In March 1989, two scientists at the University of Utah called a press conference to announce that they had discovered the holy grail of energy—nuclear fusion of atoms. Prior to their announcement, scientists had believed that to fuse two atoms of hydrogen into one of helium, which would release vast quantities of energy, would take a gigantic instrument costing billions of dollars to heat atoms to the point where fusion began. Remember that like charges repel: the forces keeping two hydrogen atoms apart are enormous at the scale of nuclei, requiring extremely high temperatures to overcome. But the Utah scientists said they had achieved fusion on their lab bench with inexpensive equipment at room temperature. Scientists were willing to at least consider the idea, in part because only three years before, a group had discovered high-temperature superconductivity, which no theory could explain. But other scientists immediately replicated high-temperature superconductivity, which is now an observational fact.

No sooner had the Utah press conference ended than scientists around the world began to try to replicate cold fusion. A few said they had, but most failed. As time went by, scientists began to retract the positive claims. Only two months after the original announcement, the American Physical Society convened a session on cold fusion at which eight speakers pronounced it dead on arrival and the ninth abstained. More than twenty years later, no one has yet replicated cold fusion.

Science is self-correcting. If a journal rejects a correct and important paper, committing an error of omission, some other journal will publish the paper eventually, as with the Krebs Cycle. When a journal publishes a flawed paper, an error of commission, other scientists will uncover the errors when they are unable to replicate the results. They will then announce their failure and the original authors will retract the paper, or become outcasts.

Thus the true test of scientific consensus is not signatures on a petition, not what anyone professes or writes on a blog, but whether a significant number of peer-reviewed papers express substantial doubt about the consensus position. Geologist and historian of science Naomi Oreskes of the University of California at San Diego set out to discover whether such papers exist for global warming. In 2005 she searched a database of all peer-reviewed scientific articles published between 1993 and 2003 for the phrase "global climate change." She found 928 that met her search criteria, read the abstracts of each, and counted and classified the articles. Seventy-five percent implicitly or explicitly agreed that the earth is warming and that human-produced carbon dioxide emissions are the cause; 25 percent dealt with research methods or ancient climates and did not take a position on global warming. How many articles presented evidence that humans are *not* causing global warming? Zero.

The deniers wasted no time in attacking Oreskes and her findings, for the stronger the evidence of consensus, the weaker their case. The first to challenge her was Benny Peiser, a senior lecturer in the School of Sport and Exercise Sciences at John Moores University in Liverpool. Peiser repeated Oreskes's analysis and claimed that he counted thirty-four articles that "reject or doubt the view that human activities are the main drivers of 'the observed warming over the last 50 years.'" Oreskes had not looked for expressions of "doubt," which could have merely reflected the natural caution of scientists.

Peiser submitted his results in a letter to *Science*, which declined to publish because, it said, the information in his letter had already been "widely dispersed on the Internet."[1] After back-and-forth criticism and challenge, Peiser began to retreat, finally admitting that

> Only [a] few abstracts explicitly reject or doubt the AGW (anthropogenic global warming) consensus which is why I have publicly withdrawn this point of my critique. I do not think anyone is questioning that we are in a period of global warming. Neither do I doubt that the overwhelming majority of climatologists is agreed that the current warming period is mostly due to human impact.[2]

Finally, Peiser acknowledged that only one article of the thirty-four met his definition of "reject or doubt," and it was not a research article but an overview by a committee of the American Association of Petroleum Geologists which had appeared in the association's own journal.

Enter the Viscount Monckton of Brenchley, former policy adviser to Margaret Thatcher. In July 2007, Monckton published a report under the auspices of the Science and Public Policy Institute, a "nonprofit institute of research and education dedicated to sound public policy based on sound science." Monckton titled his report, "'Consensus'? What 'Consensus'? Among Climate Scientists, the Debate Is Not Over."[3] In it he labeled the U.S. National Academy of Sciences a "political pressure group," reserving special venom for the Royal Society, "one of Britain's oldest taxpayer-funded lobby-groups." Oreskes had "no qualifications in climatology," Monckton said, a description that he evidently forgot applied as much to him as to her. After all, Oreskes does have a PhD in geology and is a published historian of science, whereas Monckton has no scientific credentials or publications.

Monckton claimed to have identified five articles (0.53 percent of Oreskes's 928) that she should have counted as disagreeing with the consensus view.[4] One of the five noted that global temperature had correlated with solar activity, lending credence to the claim that global warming is natural, not man-made, but the correlation ended just as the sharp temperature rise of the last three decades began. Another paper also dealt with solar variations, but on close reading turned out to corroborate the consensus position. A third, the product of an Ad Hoc Committee, had not been peer-reviewed. Number four was a review that did not include any new data; her study had not included review articles. The fifth was also a review and located in the Social Science Index, which Oreskes did not search. Oreskes had taken Monckton's wicket.

Klaus-Martin Schulte, an endocrinologist at King's College Hospital in London, challenged Oreskes in a 3,000-word open letter to her and the chancellor of her institution. Like Monckton, Schulte claimed to have found papers denying a consensus on global warming that Oreskes had missed. However, all his citations came from Peiser's letter, to which Schulte gave no attribution. Schulte's opus even repeated some of Peiser's original mistakes. A detailed comparison of Schulte's letter and Monckton's report revealed long passages that were identical.[5]

Thus the attempt by Peiser, Monckton, and Schulte to discredit Oreskes failed and her conclusion stands: "This analysis shows that scientists publishing in the peer-reviewed literature agree with IPCC, the National Academy of Sciences, and the public statements of their professional societies. Politicians, economists, journalists, and others may have the impression of confusion, disagreement, or discord among climate scientists, but that impression is incorrect."[6]

Newton's Second Law

Even if the lack of peer-reviewed articles, the ultimate test, were not enough to confirm consensus, much other evidence does confirm it. Editors and administrators with an overview of the state of science have no doubt. Donald Kennedy, former president of Stanford University and former editor of *Science*, which publishes dozens of science articles each week, said, "Consensus as strong as the one that has developed around this topic is rare in science."[7] D. James Baker, former head of the National Oceanographic and Atmospheric Administration, the chief government repository of climate information, put it this way: "There's a better scientific consensus on this than on any issue I know—except maybe Newton's second law of dynamics."[8]

The IPCC itself is a consensus organization, comprising several thousand scientists and hundreds of government officials from scores of nations. The IPCC does not do research, but surveys and summarizes the scientific literature. Thus whatever the IPCC reports must be close to the current scientific consensus—unless the IPCC is corrupt, which the deniers imply and even say it is.

As noted above, in 2001 the new Bush administration asked the National Academy of Sciences to assess the evidence for global warming. The Academy confirmed that global warming is real: "Greenhouse gases are accumulating in Earth's atmosphere as a result of human activities, causing surface air temperatures and subsurface ocean temperatures to rise."[9]

The purpose of the Kyoto Protocol, signed by 183 nations though not the United States, is "stabilization of greenhouse gas concentrations in the atmosphere at a level that would prevent dangerous anthropogenic interference with the climate system." The 183 signatory nations evidently share a consensus that accumulating greenhouse gases represent a danger.

Most scientists and the organizations that represent them stay out of politics. Now and again, an issue of science has such serious policy implications that a scientific society decides it must speak out. The science academies of thirty-three countries, from Australia to Zimbabwe, have issued statements accepting global warming and, in most cases, warning of its dangers (see Appendix). So have nearly seventy national and international science societies, including the American Association for the Advancement of Science, the American Chemical Society, the American Geophysical Union, the American Medical Association, the American Meteorological Society, the American Physical Society, the European Academy of Sciences and Art, the European Science Foundation, the Geological Society of America, the Network of African Science Academies, the Royal Societies of the UK and of New Zealand, the World Meteorological Society, and on and on. Even the American Association of Petroleum Geologists, whose members depend for their livelihood on fossil fuel combustion, in 2007 revised its statement to read: "Although the AAPG membership is divided on the degree of influence that anthropogenic CO_2 has on recent and potential global temperature increases, the AAPG believes that expansion of scientific climate research into the basic controls on climate is important. AAPG supports reducing emissions from fossil fuel use as a worthy goal."[10] *No national science academy or international organization of scientists denies the truth of global warming.*

In May 2009, two other scientific organizations spoke out. A group of Nobel Prize winners, meeting in London, hearkened back to the manifesto by Bertrand Russell and Albert Einstein in 1957, when "united scientists of all political persuasions [gathered at Pugwash, Nova Scotia] to discuss the threat posed to civilization by the advent of thermonuclear weapons." The Nobelists said that global warming "represents a threat of similar proportions."[11]

The same month, the science academies of the G8 nations and five others declared, "It is essential that world leaders agree on the emissions reductions needed to combat negative consequences of anthropogenic global warming," then urged governments to adopt targets so that, by 2050, global emissions would be 50 percent below those of 1990.[12]

In May 2010, the National Academy of Sciences issued a 409-page report titled, *Advancing the Science of Climate Change*.[13] It opened with this statement: "The body of science makes a compelling case that climate change is occurring and suggests that it threatens not just the environment and ecosystems of the world but the well-being of people today and in future generations."

To understand that scientific organizations do not make policy statements lightly, consider the example of the American Geophysical Union. As described on the organization's website, the AGU has a careful and cautious procedure for advocacy on public policy issues: "As a scientific society, AGU should not take or advocate public positions on judgmental issues that extend beyond the range of available geophysical data or recognized norms of legitimate scientific debate. Public positions adopted by AGU and statements issued on its behalf must be based on sound scientific issues and should reflect the interests of the Union as a whole."[14] The AGU does not put policy statements to a vote of its many members, just as the U.S. Congress does not put legislation to a direct vote of the people. Instead, a member-elected council acts as the AGU's legislature. For 2009–2010, the council comprises the five officers of the AGU itself, eleven Section Presidents, and eleven Section Presidents-Elect, for a total of twenty-seven.

The AGU statement, titled "Human Influence on Climate," reads in part,

> The Earth's climate is now clearly out of balance. The cause . . . is tied to energy use and runs through modern society. Warming greater than 2°C above 19th century levels is projected to be disruptive . . . and—if sustained over centuries—[will result in] melting much of the Greenland ice sheet with ensuing rise in sea level of several meters. . . . [N]et annual emissions of CO_2 must be reduced by more than 50 percent within this century.[15]

Polling Scientists

Science does not operate by opinion poll, but since we are trying to assess the deniers' claim that there is no scientific consensus, it is worth seeing what the polls have to say. Prior to 1990, few scientists outside climatology had thought much about global warming. In 1991 a survey conducted by the Gallup organization for the Center for Science, Technology, and Media sampled 400 members of the American Geophysical Union and the American Meteorological Society, the organizations with the highest percentage of climate scientists as members. In a press release, Gallup said that "67 percent of those scientists directly involved in global climate research say human-induced warming is now occurring," while "only 11 percent said that such warming was not occurring," the remainder being undecided.[16] But in an attack on Al Gore and what *Washington Post* pundit George Will said was Gore's "wastebasket-worthy" new book, *Earth in the Balance*, Will maintained that "Gore knows, or should know before pontificating, that a recent Gallup Poll of scientists concerned with global climate research shows that 53 percent do not believe warming has occurred, and another 30 percent are uncertain."[17] Gallup had to correct Will's mistaken claim, saying, "Most scientists involved in research in this area believe that human-induced global warming is occurring now."[18]

By 2007, scientific opinion about global warming had firmed. Harris Interactive surveyed nearly 500 randomly selected members of the same two organizations, the American Geophysical Union and the American Meteorological Society. The survey found that 97 percent agree that over the past 100 years the earth has warmed; 84 percent agree that human-induced warming is occurring and that global warming poses a moderate to very great danger.

In 2009, pollsters asked 10,257 Earth scientists about global warming and got 3,146 replies. Of the climatologists active in research who responded, 96 percent agreed that global temperatures have risen compared to pre-1800 levels, and 97.4 percent identified human activity as the principal cause. Of all respondents, 90 percent said that temperatures have risen compared to pre-1800 levels, and 80 percent agreed that humans significantly influence global temperatures. Fewer meteorologists and petroleum geologists were convinced, with 64 percent and 47 percent, respectively, agreeing that humans are causing global warming. Still, it is to their credit that nearly half of petroleum geologists accept global warming. The survey summed up, "It seems that the debate on the authenticity of global warming and the role played by human activity is largely nonexistent among those who understand the nuances and scientific basis of long-term climate processes."[19]

Were any doubt left that a scientific consensus on global warming exists, two publications in the first half of 2010 should remove it. In a letter to *Science* on May 7, 2010, 255 members of the U.S. National Academy of Sciences, including eleven Nobel Prize winners, wrote:

(i) The planet is warming due to increased concentrations of heat-trapping gases in our atmosphere.

(ii) Most of the increase in the concentration of these gases over the last century is due to human activities, especially the burning of fossil fuels and deforestation.

(iii) Natural causes always play a role in changing Earth's climate, but are now being overwhelmed by human-induced changes.

(iv) Warming the planet will cause many other climatic patterns to change at speeds unprecedented in modern times, including increasing rates of sea-level rise and alterations in the hydrologic cycle. Rising concentrations of carbon dioxide are making the oceans more acidic.

(v) The combination of these complex climate changes threatens coastal communities and cities, our food and water supplies, marine and freshwater ecosystems, forests, high mountain environments, and far more.[20]

How can one interpret the scientists' statement? There are only three possibilities: (1) the world's best scientists are simply wrong about a matter within their expertise and those without such expertise are right; (2) 255 members of the National Academy of Sciences are part of a global conspiracy; (3) the Academy members are right.

In late June 2010, three authors published an article in the *Proceedings of the National Academy of Sciences* titled, "Expert Credibility in Climate Change."[21] They used "a dataset of 1,372 climate researchers and their publication and citation data" to show that among the most active publishers on climate, 97–98 percent "support the tenets" of global warming. The study also found that "the relative climate expertise and scientific prominence of the researchers unconvinced of anthropogenic climate change are substantially below that of the convinced researchers."

An Extremely Pernicious Development

While some deniers argue that there is no consensus on global warming, others take a different tack, implicitly accepting that a consensus exists, then denouncing consensus itself as false and malicious. That is the strategy of two skilled rhetoricians: John Tierney, *New York Times* columnist, and late author Michael Crichton. Both allege that the mere existence of a scientific consensus is *prima facie* evidence of falsity.

Yale grad Tierney takes consistently conservative or libertarian positions, writing from "TierneyLab," where, he claims, he "check[s] out new research and rethink[s] conventional wisdom about science and society." Tierney has a long anti-environmental history. He denounced Rachel Carson's seminal book, *Silent Spring*, as "a hodgepodge of science and junk science." In 1995 he said the evidence that secondhand smoke is harmful was "dubious."[22] In a June 30, 1996, article in the *New York Times Magazine* titled "Recycling is Garbage," Tierney wrote, "Recycling may be the most wasteful activity in modern America: a waste of time and money and natural resources." After President Obama nominated Harvard physicist John Holdren as his Science Advisor, Tierney said, "Dr. Holdren is certainly entitled to his views, but what concerns me is his tendency to conflate the science of climate change with prescriptions to cut greenhouse emissions."[23] Does this mean that the proprietor of "TierneyLab" denies the reality of the greenhouse effect?

Maybe, because the founding principles of TierneyLab are: "1. Just because an idea appeals to a lot of people doesn't mean it's wrong. 2. But that's a good working theory."[24] Carrying denial to a *reductio ad absurdum*, Tierney implies that the more who accept an idea, the more apt it is to be false. This belies the very notion of progress in human affairs and in science, as well as the need for knowledge to accumulate and approach the truth. According to Tierney, the only ideas likely to be correct are those that the mainstream rejects. Tierney segues from the obvious truism that a consensus is not always right to the conclusion that it is wise to assume it is wrong. Were the science correspondent for the *New York Times* correct, science would have brought no progress and the world would be stuck in the Dark Ages.

Author Michael Crichton died of lung cancer in 2008. He had become a multimillionaire using science as the backdrop to his novels, beginning with *The Andromeda Strain* and continuing through *Jurassic Park* and many other popular works. But when science conflicted with his ideology, Crichton reversed himself and wrote the anti-science polemic, *State of Fear*.

In January 2003, Crichton gave a speech at the California Institute of Technology that presaged the book. In the speech, titled, "Aliens Cause Global Warming: An historical approach detailing how over the last thirty years scientists have begun to intermingle scientific and political claims," Crichton went further than Tierney, denouncing "consensus science [as] an extremely pernicious development that ought to be stopped cold in its tracks . . . the first refuge of scoundrels." He concluded, "If it's consensus, it isn't science. If it's science, it isn't consensus. Period."[25]

In a talk that same year to the Commonwealth Club of San Francisco, titled, "Environmentalism as Religion," Crichton revealed why he is a denier. "The greatest challenge facing mankind," he said, "is the challenge of distinguishing reality from fantasy, truth from propaganda."[26] Agreed. But no one has done more to confuse fantasy about global warming with reality than Crichton. "Environmentalism seems to be the religion of choice for urban atheists," he went on. "I am thoroughly sick of politicized so-called facts that simply aren't true. It isn't that these 'facts' are exaggerations of an underlying truth. Nor is it that certain organizations are spinning their case to present it in the strongest way. Not at all—what more and more groups are doing is putting out lies, pure and simple. Falsehoods that they know to be false."

In his Cal Tech speech, Crichton denied that there was evidence that secondhand smoke causes cancer, accusing the EPA of "cheating" and "fraudulent science."[27] To the Commonwealth Club, he said "I can tell you that secondhand smoke is not a health hazard to anyone and never was, and the EPA has always known it." But the International Agency for Research on Cancer says that secondhand smoke *is* a health and cancer risk. The U.S. Surgeon General in 2006 estimated that living or working in a place where smoking is permitted increases a nonsmoker's risk of developing heart disease by 25–30 percent and lung cancer by 20–30 percent. The report also links passive smoke to sudden infant death syndrome (SIDS), respiratory problems, ear infections, and asthma attacks in children.

To support his condemnation of consensus, in the same speech Crichton said, "Consensus is invoked only in situations where the science is not solid enough. Nobody says the consensus of scientists agrees that $E = mc^2$. . . . It would never occur to anyone to speak that way."

But Einstein's equation is not a theory; scientists have shown it to be a fundamental law of physics. Such complex processes as global warming do not lend themselves to experiment and to simple but profound equations. Such processes are difficult to understand, to model, and to test. Knowledge grows slowly.

But let us probe $E = mc^2$ a bit. Einstein proposed the mass–energy equivalence in a 1905 paper titled, "Does the Inertia of a Body Depend Upon Its Energy Content?"[28] He did not present $E = mc^2$ as an established law, but put it forward and asked whether it was correct. The paper came out during a period of enormously productive ferment in physics, in what some have called Einstein's *annus mirabilis*, his "Miraculous Year" of pathbreaking papers, including those introducing special relativity and the photon theory of light. Einstein framed the title of the mass-energy paper as a question and ended it with an "if": "If the theory corresponds to the facts, radiation conveys inertia between the emitting and the absorbing bodies." No one, and certainly not Einstein, at the time claimed there was a consensus about $E = mc^2$. The notion was controversial and Planck, for one, questioned whether Einstein was correct. Consensus grew gradually. It took the discovery of the positron in 1932 to show that Einstein had unearthed a law of physics. The evolution of the great equation shows how consensus builds and demonstrates the exact opposite of what Crichton claimed. Consensus is how we create knowledge.

Crichton sums up: "Whenever you hear the consensus of scientists agrees on something or other, reach for your wallet, because you're being had."[29] You might be better off to reach for that wallet when skilled writers like Crichton, with their easily spotted agendas, try to dupe you into believing that a consensus *must* be wrong.

Consensus is a natural stage in the evolution of human thought. Naomi Oreskes put it this way: "Scientific knowledge is the intellectual and social consensus of affiliated experts based on the weight of available empirical evidence, and evaluated according to accepted methodologies." She continued, "If we feel that a policy question deserves to be informed by scientific knowledge, then we have no choice but to ask, what is the consensus of experts on this matter?"[30]

Discovery of Global Warming

We have established that scientists share a consensus on global warming, but why do they? What is the theory of global warming and what is the evidence for it? This chapter and the next two provide the answers.

Most people likely heard the term *global warming* for the first time in the 1990s and may be unaware that it is one of the most venerable ideas in science.[1] In 1896, using only pencil and paper, a Swedish chemist and eventual Nobel Prize winner named Svante Arrhenius calculated that if the amount of carbon dioxide gas in the atmosphere were to double, global temperatures would rise 5–6°C (9–11°F). In 2001 the Intergovernmental Panel on Climate Change, a large, modern, international team of scientists, had to revise Arrhenius's estimate only slightly, concluding that doubling atmospheric carbon dioxide would raise temperatures 2.5–5.8°C (3–8°F).

One of the Oldest Theories

The origins of our understanding that gases in the atmosphere influence climate go back well before Arrhenius. In 1824 a French polymath named Joseph Fourier recognized that atmospheric gases trap heat, raising the surface temperature enough to allow us to inhabit the planet. The core principle of the greenhouse effect is older than all but a few extant scientific theories.

Once chemists and physicists were able to identify and isolate the major gases that comprise the air, an obvious question was which ones do the actual absorbing of radiation that Fourier had discovered. An Irishman named John Tyndall devised an apparatus that allowed him to pass heat through individual atmospheric gases one at a time—nitrogen, oxygen, water vapor, carbon dioxide, and hydrocarbons—and measure how much each absorbed. Tyndall found that, collectively, atmospheric gases absorbed enough radiation to have caused "all the mutations of climate which the researches of geologists reveal."[2] Tyndall's reference to geology revealed that his interest was not in forecasting future climates, but in trying to explain past ones, especially the solution to a mystery that scientists had only recently discovered: in the Northern Hemisphere, not once, but at least four times, huge sheets of ice as much as a mile thick had advanced southward, obliterating everything in their path. After each advance the ice retreated, only to advance again.

Tyndall's research showed that short-wavelength radiation from the sun passes largely unimpeded through the atmosphere to strike and warm the earth's surface. But the radiation escaping the surface lies in the infrared, long-wavelength band, which atmospheric gases like water and carbon dioxide strongly absorb. Some of this absorbed radiation is reradiated back down to the surface, warming it enough to allow us to live here. Scientists came to call the phenomenon the "greenhouse effect" (fig. 4.1), although in a real greenhouse, the glass merely blocks convecting heat from escaping.

Tyndall's discovery meant that if the amounts of different gases in the atmosphere were to change, allowing the atmosphere to absorb more

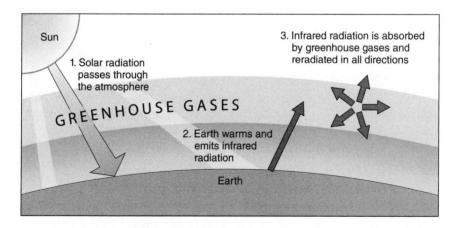

Figure 4.1 The greenhouse effect (Courtesy of John Cook)

radiation at one time and less at another, climate might also change. That possibility intrigued Arrhenius, a child prodigy and scientific genius who learned to read at age three and skipped the first four grades. His PhD dissertation on electrolytic conductivity was so novel that his professors could not understand it and gave Arrhenius a fourth-class degree, preventing him from getting a job teaching chemistry at any Swedish university.[3]

Like Tyndall and many other nineteenth-century scientists, Arrhenius was curious about the cause of the ice ages. Scientists recognized that we happen to live in a warm, relatively ice-free "interglacial" period, raising the question of whether and when the obliterating ice might again start to advance. Figuring out the cause of the ice ages was more than an academic question.

Drawing on the work of Tyndall, Arrhenius began to investigate the role of carbon dioxide gas in the atmosphere. He cleverly used observations of the Moon to determine how much radiation water vapor and carbon dioxide in the atmosphere absorb. After months of laborious calculations, in 1896 Arrhenius reported that if the amount of carbon dioxide in the atmosphere were cut in half, temperatures would drop by 4–5° C (7–9° F). Enough, perhaps, to start an ice age. Conversely, as already noted, were atmospheric carbon dioxide to double, temperatures would rise by 5–6° C (9–11° F), a "climate sensitivity" close to today's estimate.

Arrhenius presented his findings in 1908 in a popular book called *Worlds in the Making*. By then, world coal production had risen to about 900 million tons annually, leading him to perceive that, "Although the sea, by absorbing carbon dioxide, acts as a regulator of huge capacity, which takes up about five-sixths of the produced carbonic acid . . . the slight percentage of carbonic acid in the atmosphere may by the advances of industry be changed to a noticeable degree in the course of a few centuries."[4] But, as might be expected from a native of Sweden, Arrhenius concluded that more carbon dioxide and a warmer Earth might be a good thing: "We may hope to enjoy ages with more equable and better climates, especially as regards the colder regions of the earth, ages when the earth will bring forth much more abundant crops than at present, for the benefit of rapidly propagating mankind."[5] In any case, such benefits lay far in the future, for Arrhenius presumed that it would take thousands of years for carbon dioxide to increase appreciably.

Arrhenius merely showed that carbon dioxide *could* affect global temperatures, not that it actually had.[6] Most scientists of his day believed that the oceans would absorb enough of any excess carbon dioxide to prevent it from building up in the atmosphere. Tyndall had found that water vapor also strongly absorbs infrared radiation; there is a lot more water vapor in the atmosphere than carbon dioxide. Perhaps water vapor absorbs so much radiation that it leaves none for the minute traces of carbon dioxide to affect. For these and other reasons, scientists at the turn of the century saw little cause to pursue Arrhenius's findings.

This was far from the only time that an insightful scientist, years or decades ahead of others, proposed a theory so novel that that there was no obvious experiment to corroborate or falsify it. With no way to test the greenhouse effect, scientists turned to research apt to be more productive. But they had not falsified Arrhenius's theory. Books and scientific papers continued to mention him, no doubt prompting some young scientists, always ready to question authority, and others of contrarian tendencies, to read what their long-ago predecessors had written. Even though it appeared forgotten, the greenhouse effect awaited rediscovery in the dusty corners of science libraries.

One who became interested in Arrhenius's ideas in the 1930s was a British engineer and specialist in steam production named Guy Stewart Callendar. He proved as indefatigable as Arrhenius. In a presentation to the Royal Society and in a 1938 article, Callendar reported that he had used the temperature observations made at 200 meteorological stations to show that the earth was getting hotter: 0.009°F per year according to his calculations. Why? Because burning of coal had added "about 150,000 million tons of carbon dioxide to the air during the past half century." That much carbon dioxide, according to Callendar, should have caused a temperature increase of 0.005°F per year, not far below the observed increase.[7] Callendar reported that in 1900 the atmosphere had held 290 parts per million of carbon dioxide by volume, remarkably close to modern estimates.

Like Arrhenius, Callendar saw advantages in a warmer Earth: "The combustion of fossil fuel is likely to be beneficial to mankind in several ways, besides the provision of heat and power," he wrote, including "the growth of favorably situated plants [which is] directly proportional to the carbon dioxide pressure." Callendar may also have glimpsed the danger: "As regards the reserves of fuel these would be sufficient to give at least ten times as much carbon dioxide as there is in air at present."[8]

A Large-Scale Geophysical Experiment

Climate science, like other branches, benefited from the hundreds of millions of dollars poured into science and technology during World War II. Scientists invented a whole new set of instruments and, once the war was over, used them for peacetime research. Then came the Cold War to prompt even more generous funding of some areas.

One objection to the greenhouse theory was the old notion that most of the Earth's carbon dioxide dissolves in the oceans. It seemed reasonable that if humans were to increase the amount in the atmosphere slightly, the oceans would merely absorb the small excess and all would be as before. Scientists needed a way to trace carbon through the atmosphere and oceans to learn where the carbon released by the combustion of fossil fuels had wound up.

World War II atomic research provided the method. Willard Libby of the University of Chicago invented the radiocarbon dating technique, which revolutionized anthropology and archaeology and earned Libby the Nobel Prize in Chemistry. Within a few years, scientists at the Scripps Institute of Oceanography in La Jolla, California, led by its director, Roger Revelle, used radiocarbon as a tracer to show that the atmosphere does contain fossil carbon, which could only have gotten there through the burning of fossil fuels. The research also revealed that a typical molecule of carbon dioxide remains in the atmosphere for only about a decade before the oceans absorb it. But another question was critical to understanding the greenhouse effect: once that molecule entered the ocean, how long would it remain there?

Revelle, a student of ocean chemistry, soon realized that chemical reactions buffered the absorption of carbon dioxide by sea water. The buffering meant that the oceans could absorb only about 10 percent as much carbon dioxide as they could have in its absence. As historian of science Spencer Weart puts it in his indispensable account, "Although sea water did rapidly absorb carbon dioxide, most of the added gas would promptly evaporate back into the air before the slow oceanic circulation swept it into the abyss."[9] This meant that humanity could not depend on the oceans to absorb an unlimited amount of carbon dioxide, one of the most important and ominous discoveries of modern science.

In a classic 1957 paper, Revelle became one of the first to grasp the import of the greenhouse effect: "Human beings are now carrying out a large scale geophysical experiment of a kind that could not have happened in the past nor be reproduced in the future."[10] But even Revelle, likely the most knowledgeable person on the subject, did not see the experiment as particularly threatening. In the 1950s, scientists did not know how much carbon dioxide the atmosphere held, much less whether the amount was increasing.

Better instruments soon became available to measure atmospheric carbon dioxide concentrations, but it was a tricky business, as air blown from factories, highways, and the like could throw off the results. A student at the California Institute of Technology, Charles David Keeling, began to measure atmospheric carbon dioxide in air and learned how to avoid the contamination that plagued the measurements. Revelle

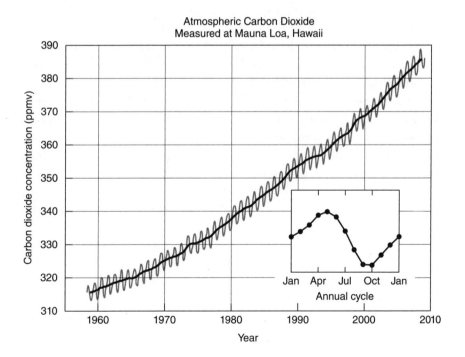

Figure 4.2 The Keeling Curve

persuaded Keeling to join him at Scripps and managed to secure funding to allow the young scientist to set up more accurate instruments to measure total carbon dioxide concentration at two isolated sites far from contaminating sources: the top of Mauna Loa volcano on Hawaii and a research station in Antarctica. After only two years of measurements, Keeling's Antarctic data showed an increase in atmospheric carbon dioxide concentration. Thus not only had Keeling figured out how to overcome contamination to detect minute differences in atmospheric carbon dioxide levels, he had found that the amount was increasing, just as Arrhenius had predicted. But there was one major difference: Arrhenius, Callendar, and others had thought it would take hundreds or thousands of years for carbon dioxide to build up appreciably; instead it was rising yearly. This raised the questions of how long CO_2 had been rising, how long might it continue to rise, and what would be the consequences for humanity?

Keeling's station on Antarctica had to close for lack of funding, but although his Mauna Loa Station nearly ran out of money several times, it managed to keep operating and still does today. Funds for Keeling's research may be some of the best spent in history, for his results have continued to show a steady and seemingly inexorable rise in atmospheric carbon dioxide. In no year since Keeling began his measurements have carbon dioxide concentrations declined (fig. 4.2).

Thus by 1960 the greenhouse effect had evolved from theory to observational fact to dimly perceived threat. Keeling's punctilious research showed that it was possible to measure carbon dioxide concentrations accurately and that they were rising. First a few individuals, and then panels of scientists, began to perceive a danger.

The Greenhouse Effect

From Curiosity to Threat

How fast and how far could atmospheric carbon dioxide rise and with what effects? One of the first groups to take up those questions was the President's Science Advisory Committee (PSAC), which Dwight D. Eisenhower had upgraded from President Truman's Science Advisory Committee and relocated to the White House. (Today we know it as the President's Council of Advisers on Science and Technology, or PCAST). In 1965 the Environmental Pollution Panel, a subcommittee of PSAC that included Roger Revelle and Charles David Keeling, reported that "by the year 2000 there will be about 25% more CO_2 in our atmosphere than at present," warning that the increase "will modify the heat balance of the atmosphere to such an extent that marked changes in climate . . . could occur."[1]

President Lyndon Johnson became the first U.S. president to caution about the possible consequences of the greenhouse effect, saying in a special message to Congress in February 1965: "This generation has altered the composition of the atmosphere on a global scale through . . . a steady increase in carbon dioxide from the burning of fossil fuels."[2] (Almost forty years later, in an interview with *People* magazine on the occasion of his sixtieth birthday in July 2006, President George W. Bush was unwilling to go as far as Johnson, saying only, "I think there is a debate about whether [global warming is] caused by mankind or whether it's caused naturally, but it's a worthy debate. It's a debate, actually, that I'm in the process of solving.")[3]

In response to the 1965 PSAC report, the National Academy of Sciences conducted its own review. Displaying the natural caution of scientists and the tendency of committees to understate, the Academy panel concluded that although past changes in climate had locally catastrophic effects, "they did not stop the steady advance of civilization."[4] Greenhouse warming was real, but not particularly threatening.

Models Predict Warming

In 1967 one of the earliest computer models of climate addressed the question that Arrhenius had posed: if atmospheric carbon dioxide concentrations were to double, how much would global temperature rise? The model, primitive by today's standards, pegged the "climate sensitivity" at 2°C (3.6°F), lower than the Swedish Nobelist had calculated 71 years earlier, but not by much. By 1975 an improved computer model raised the estimate to 3.5°C (6.3°F). The exact figure was less important than confirmation that adding carbon dioxide to the atmosphere by burning fossil fuels would cause a significant temperature increase. Thanks—or no thanks—to the amplifying climate feedbacks, a tiny amount of carbon dioxide can make a disproportionate difference to global temperatures.

In 1977 the President's science adviser, geophysicist Frank Press, asked the National Academy of Sciences to review the results of computerized climate modeling and determine whether the models were reliable. Since these early models took weeks to run and even then did not settle down to an equilibrium climate for the earth, it was a reasonable and important question.

In the chair of the Academy committee was Jule Charney, a distinguished climate modeler from MIT known as the "Father of Dynamical Meteorology." The Academy asked the committee this question: "If we were certain that atmospheric carbon dioxide would increase on a known schedule, how well could we project the climatic consequences?"[5] The underlying assumption of the charge was that "atmospheric concentrations of carbon dioxide are steadily increasing, and these changes are linked with man's use of fossil fuels and exploitation of the land" (vii). In its report, the Charney committee noted that "the more realistic of the modeling efforts predict a global surface warming of between 2°C (3.6°F) and 3.5°C (6.3°F)." With scientific understatement aided by a double negative, the report concluded, "It is significant that none of the model calculations predicts negligible warming" (1).

Charney's panel paid special attention to feedbacks, which can either reinforce an effect or diminish it. Scientists had known since the nineteenth century of positive, or amplifying, climate feedbacks, the ice-water feedback being one of the most important. Ice reflects most incoming sunlight; sea water absorbs most of it. When sea ice melts, the resulting ocean water absorbs more solar radiation and warms, which melts more ice, which causes the ocean to absorb more sunlight, and so on. (Since positive climate feedbacks tend to have negative consequences for us, I refer to them as *amplifying feedbacks*.) An even stronger amplifying feedback derives from water itself. As temperature rises, more water evaporates, but a warmer atmosphere holds more water vapor. And, as Tyndall found, water is a strong greenhouse gas. The extra water vapor in the atmosphere absorbs more heat, which causes more warming and evaporation, and so on. Mankind can hope that there are negative climate feedbacks to offset the effects of rising atmospheric carbon dioxide, but the Charney committee could find none large enough to make a difference, writing, "We have examined with care all known negative feedback mechanisms. . . . We have tried but have been unable to find any overlooked or underestimated physical effects that could reduce the currently estimated global warmings due to a doubling of atmospheric carbon dioxide to negligible proportions or reverse them altogether" (2–3). As early as the mid-1970s, scientists sought negative climate feedbacks, but in vain.

After reviewing the result of global climate modeling, the Charney committee estimated the increase from doubling atmospheric carbon dioxide at 3°C (5.4°F), the same sensitivity that James Hansen described as "nailed" three decades later in his 2008 AGU talk (17). The Charney committee found that as atmospheric carbon dioxide levels increased, "Warming will eventually occur . . . and the associated regional climatic changes . . . may well be significant" (3). But the committee declined to speculate about the potential socioeconomic consequences. The Charney report offered a clear warning and, following the universal rule of scientific panels, thoroughly documented by Spencer Weart, called for more research funds.

Globetrotters

In 1979 the World Meteorological Organization and the United Nations Environmental Program sponsored a World Climate Conference in Geneva, Switzerland. The keynote speaker summed up the state of knowledge: "The potential consequences of increasing atmospheric CO_2 resulting from fossil fuel combustion are already a major concern. . . . The implications of further projected increases are uncertain, but the weight of scientific evidence predicts a significant global surface temperature increase." The conference called on nations "to foresee and prevent potential man-made changes in climate that might be adverse to the well-being of humanity."[6]

International climate meetings had begun to proliferate. A sample of those from the late 1980s would include the Villach Conference (October 1985), the Toronto Conference (June 1988), the Ottawa Conference (February 1989), the Tata Conference (February 1989), the Hague Conference (March 1989), the Noordwijk Ministerial Conference (November 1989), the Cairo Compact (December 1989), and the Bergen Conference (May 1990), not to mention a host of meetings of chemists, physicists, meteorologists, and others at which future climates were discussed. Scientists studying a global problem had become globetrotters.

By now, the conservative administration of U.S. President Ronald Reagan had begun to worry that action to reduce carbon dioxide emissions could harm the American economy. Reagan had surprised a lot of people, including many of his conservative friends, by signing the Montreal Protocol to eliminate chlorofluorocarbons like Freon, which scientists had shown were depleting the atmospheric ozone layer that protects us from damaging ultraviolet radiation. (CFCs are also potent greenhouse gases.) But only a few chemical companies made CFCs, and less dangerous substitutes were readily available. In contrast to the expendable CFCs, the U.S. economy runs on coal, oil, and natural gas. Growing alarm over carbon dioxide emissions from burning fossil fuels might lead to a Montreal-like protocol to reduce carbon emissions, allegedly crippling the economy—and on Reagan's pro-business watch. To have scientists meeting where and when they liked, saying whatever they pleased, issuing disquieting statements, could force governments to respond. The solution was to create a new, international scientific body and ensure that government representatives vetted its reports.

The U.S. signed off on a proposal from the United Nations to create an overarching climate advisory committee called the Intergovernmental Panel on Climate Change, mandated to "provide the decision-makers and others interested in climate change with an objective source of information."[7] Governments would appoint their own scientists to the panel. Diplomats and government bureaucrats from scores of nations would oversee the scientists' work and edit their reports. The structure guaranteed that the IPCC reports would neither appear too rapidly nor overdramatize the extent of global warming. From the get-go, by design, the IPCC was a conservative organization predestined for understatement.

The Intergovernmental Panel on Climate Change

The first IPCC report, appearing in 1990, confirmed that the earth was warming, though the panel could not say what caused the warming. Variations in the Sun's intensity offered a possible cause in theory, but the IPCC ruled out that explanation. The panel's most important finding was to confirm the conclusion of scientists from Arrhenius on: if atmospheric carbon dioxide levels continued to rise, global temperatures could increase by several degrees.

The 1990 report served as backdrop for the first Earth Summit, held in Rio de Janeiro in 1992. President George H. W. Bush attended but would agree to sign only the most timorous treaty. Nevertheless, he did urge world leaders to take "concrete action to protect the planet."[8]

Rio was the first but far from the last international accord to urge, but not to require, that nations reduce greenhouse gas emissions. Since at the time the United States caused more than one-third of global carbon dioxide emissions, as long as America refused to accept mandatory reductions there was no point in other nations doing so. After paying lip service to voluntary reductions, the conferees left Rio to return to business as usual.

Meanwhile, teams of scientists continued to perfect their computerized climate models. Since the 1970s, modelers had taken advantage of Moore's Law—computing power at constant cost doubles every eighteen months—to become ever more sophisticated. They adjusted their models by "hindcasting": seeing how well they could replicate known past temperatures. By the early 1990s, scientists had found that when their computer models used only natural factors, such as variation in the intensity of the Sun's rays, the amount of aerosols in the atmosphere, volcanic dust, and the like, the models failed to replicate past temperatures. Only when the modelers added the effects of greenhouse gas emissions to the natural causes did their models reproduce twentieth-century temperatures.

In its Second Assessment Report, issued in 1995, the IPCC surveyed the science of global warming, concluding that, "*The balance of evidence suggests a discernible human influence on global climate.*" The statement was as neutral as possible, qualified by "balance of evidence," "suggest," "discernible," and "influence"—not the strongest words. But the cautious wording could not mask the fundamental conclusion: having established that global warming is real, the global scientific community was becoming increasingly certain that humans were the cause.

The 1995 report served as background for the next climate summit, held in Kyoto, Japan, in 1997. Over 6,000 official delegates attended. The United States delegation urged that the main greenhouse gas producers pledge to lower their emissions to the 1990 level, but argued that they should do so voluntarily. Led by China, the developing nations balked, insisting that it was hypocritical of developed nations to have polluted their way to First World status and then to deny the developing countries the same opportunity by the same path. The Kyoto meeting nearly collapsed, rescued only by Vice President Al Gore, who flew in on the last day to engineer a compromise agreement. The resulting Kyoto Protocol was pitifully weak, requiring no nation to do anything and exempting the developing countries—the greenhouse gas emitters of tomorrow—from even voluntary reductions. Gore's own nation never signed.

The Kyoto Protocol was supposed to be in effect only for a few years, after which nations would adopt a stronger one. Nearly all the developed nations failed to meet their voluntary targets and the undeveloped ones had no targets. History will have to judge whether the tiny foothold the Kyoto Protocol offered turned out to make a difference. The protocol at least showed that an international climate agreement is possible, though it represented only baby steps.

By 2001 more research on the effects of global warming had become available and the climate models were more numerous and accurate. In its Third Assessment report (2001), the IPCC said it was "*likely*"— defined as having a probability of 66 to 90 percent—*that humans had caused the observed global warming.* Moreover, the report predicted that on the present path, global temperatures were likely to rise from 1.4 to 5.8°C (2.5°F to 10.4°F) by 2100. Sea level would rise from 10 to 90 centimeters (4 inches to 35 inches), the wide range reflecting the use of several climate models each with different assumptions for the growth of carbon dioxide.

The IPCC's Fourth Assessment, issued in 2007, went further, saying that the warming of Earth's climate system is *"unequivocal."* Now the IPCC had the confidence to declare it *"very likely,"* defined as more than 90 percent probable, that humans were causing global warming. The potential effects were ominous: world temperatures will rise between 1.1 and 6.4°C (2.0 and 11.5°F) by 2100, and sea levels will rise between 18 and 59 cm (7 and 23 inches). Curiously, the Fourth Assessment's estimate of potential sea level rise was lower than the Third Assessment's, suggesting that the IPCC in 2007 had come to see global warming as less dangerous than it had in 2001. But a close reading showed that the 2007 report had warned that, "Dynamical processes related to ice flow not included in current models but suggested by recent observations could increase the vulnerability of the ice sheets to warming, increasing future sea level rise."[9] The Fourth Assessment had not included some of the most recent and not-yet-published research showing that the world's white areas were losing ice faster than scientists had measured earlier. Soon the IPCC and others would begin to project significantly higher sea level rises than even in the Third Assessment.

In its 2007 report, the IPCC concluded that due to the timescales of climate processes and feedbacks, "Both past and future anthropogenic carbon dioxide emissions will continue to contribute to warming and sea level rise for more than a millennium."[10] Thus although the IPCC and climate modelers tend to tell us what the world could be like in 2100, that date is arbitrary. Whatever is going on in 2100 will continue for hundreds, perhaps thousands of years, and likely grow worse along the way. A few readers of this book, and certainly their grandchildren, will live to experience the twenty-second century for themselves. They will likely either thank today's global warming deniers, or condemn them.

As the IPCC has strengthened its conclusions over the years, the deniers, unable to refute the panel's science, have made an all-out effort to discredit the panel and its members, claiming that the IPCC has a liberal, pro-environmental bias that causes it to exaggerate the threat from global warming. But as I have already shown, that claim does not pass the smell test. Were it true, why was the IPCC so cautious in 1990? Why did it wait until 2007 to say that global warming was unequivocal and caused by humans? Remember that the IPCC was set up as a conservative organization designed to prevent scientists from being too outspoken too soon.

The greenhouse effect is one of the oldest and most thoroughly studied ideas in science. Scores of scientific panels and national and international scientific organizations confirm that global warming is real and caused by burning fossil fuels and the subsequent buildup of CO_2. Not a single panel or organization has come to the opposite conclusion. That global warming is real and caused by humans is the overwhelming consensus of the world's scientists.

Projecting the consequences of global warming into the future has to worry any thinking person. The Mauna Loa station found that carbon dioxide concentrations rose from about 315 parts per million by volume when the measurements began in 1958 to 393 ppm by June 2010. (The latter figure, for example, means that of a million gallons of air, 393 gallons are carbon dioxide.) Many scientists regard 450 ppm as the point-of-no-return, after which amplifying feedbacks will prevent us from limiting the advance of global warming. If carbon dioxide concentrations increase by 2.5 ppm annually, close to the present rate, they will reach 450 ppm in the early 2030s.

Global Warming

All You Really Need to Know in One Chart

Why do scientists agree that global warming is real and caused by humans? Because the evidence convinces them. Listen to the deniers and you would believe that the evidence for global warming is so complicated and inconclusive that even specialists rightfully disagree. Not so: the core evidence for global warming is plain and fits on a single chart (fig. 6.1).

The front frame of the chart shows how, as humans began to burn larger amounts of coal after the Industrial Revolution, carbon emissions rose sharply and have continued to do so into the Oil Age. The middle frame shows how global atmospheric carbon dioxide concentrations rose in concert with carbon emissions. The rear frame shows the concomitant rise of global temperatures (by which, to be technical for a moment, I refer to the global mean annual surface temperature).

If we knew nothing but the chart, we would know only that fossil fuel emissions, atmospheric CO_2, and temperature correlate. That would seem unlikely to represent coincidence, but we would not know which of the three factors might have caused the others. And as always, we would need to remember that correlation is not causation. But in this case we do know more: we know that burning fossil fuels has put increasing amounts of CO_2 into the atmosphere. We know from carbon isotope studies that the increased carbon dioxide in the atmosphere came from fossil fuel combustion rather than from some natural cause.[1] We know that the greenhouse effect, a fact of physics, requires that the more CO_2 in the atmosphere, the higher the temperature.

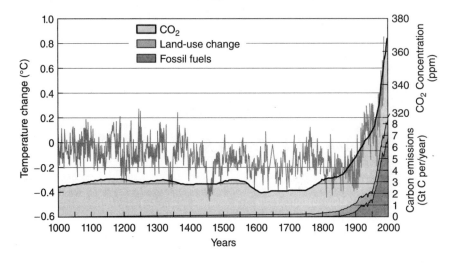

Figure 6.1 The case for global warming (Courtesy the Arctic Climate Impact Assessment; copyright ACIA, 2004)

To take the opposite view and deny that the chart shows cause and effect requires one to believe two things for which there is no evidence: first, during the twentieth century, some unknown factor suppressed the greenhouse effect, preventing the increase in carbon dioxide from causing the increase in temperature. This could only be the elusive negative climate feedback that scientists have sought in vain at least since the 1970s. Waiting for it is like refusing to take out life insurance because of a family rumor that a rich uncle, whom no one has ever met, is going to die and leave you his fortune.

Second, if for the sake of argument we grant that something did suppress the greenhouse effect, then some natural process must have caused the temperature increase. But as I will show in chapter 12, scientists have investigated each natural cause of temperature variation—the Sun's activity, atmospheric ozone, volcanic eruptions, sulfur aerosols—and none by itself or in any combination can explain the twentieth-century temperature increase. Thus the choice is to accept obvious cause and effect based on known scientific facts, or to accept two unknowns for which there is no evidence.

Moreover, instead of looking at CO_2 levels and global temperatures historically, we can simply observe the abundant, incontrovertible, and growing evidence that the earth actually is warming. The year 2010 tied 2005 as the warmest year on record. The year 2009 was the warmest on record in the Southern Hemisphere and the second warmest globally. Starting with the 1960s, global temperatures have risen with each decade. In the 1950s and 1960s more record low temperatures were set in the United States than record highs; during the first decade of this century, both the United States and Australia set twice as many record high temperatures as record lows. Nights are warming faster than days and winters faster than summers, both as predicted by global warming theory. Greenland and Antarctica are warming and losing ice, as are the world's glaciers. The volume of Arctic sea ice is declining faster than projected in the most recent IPCC models. In North America, the western snowpack is melting up to twenty days earlier in the spring. The frost season is shrinking. Northern Hemisphere snow cover is declining, as is the amount of permafrost. The oceans have warmed steadily since 1970 and grown more acidic as they dissolve more CO_2. Sea level has risen and, from 2005 to 2010, at the fastest rate on record. Plants and animals are shifting their ranges in ways predicted by global warming. And so on. As Chico Marx said in *Duck Soup*, "Who are you going to believe? Me, or your own eyes?"

Anyone with a high school physics course can explain the chart above. No theory is required; no computer model; no one's opinion; no appeals to consensus. For this simplest of charts, global warming deniers have no explanation.

The chart looks backward; we must also look forward. Scientists can only model the feedbacks that they have identified. So far, and sadly for humanity, all important climate feedbacks have turned out to be amplifying, ones that make global warming worse, while the search for important negative feedbacks continues in vain. Do we want to bet our children's future on the possibility that someone, someday, will discover enough negative feedbacks to offset the amplifying ones?

Go back and look at the chart once more. The deniers ask us to believe that fossil fuel emissions and atmospheric CO_2 can continue to rise, yet for some reason that they cannot explain, global temperatures will stop rising and fall of their own accord.

Tobacco Tactics

The Scientist-Deniers

According to a Pew Foundation poll conducted in November 2009, two-thirds of Americans believe that scientists disagree about global warming. Yet as we have seen, every scientific organization that has spoken, and the vast majority of individual scientists, accept global warming. Oreskes found not a single peer-reviewed paper that expressed doubt. What could have given the public such a misleading impression? There are several reasons, but one is that an industry of denial has mounted a successful public relations campaign, employing the same small set of apostate scientists to claim that global warming is false, or at least nothing to worry about, many of whom formerly applied their skills to denying that smoking causes lung cancer.[1] Review the list of "experts" to whom the media turn to "balance" the opinions of mainstream scientists, and the same handful of names turns up. Search Amazon.com for books that deny global warming and there they are again. Peruse the list of speakers at a Heartland Institute Conference, and ditto. Examine the names of "fellows" or "research associates" of the denier organizations—the same. This chapter shines a spotlight on a sample of the scientist-deniers and their claims, beginning with the grandfather of global warming denial, S. Fred Singer.

All of Those People Are Wrong

A Heartland Institute "expert" and conference speaker, S. Fred Singer has denied global warming for decades, speaking, writing, and testifying widely. One of the best-credentialed of the scientist-deniers, Singer has a bachelor's degree in engineering from Ohio State University and a PhD in physics from Princeton. He served as director of the U.S. National Weather Satellite Center, deputy administrator of the Environmental Protection Agency, and professor of environmental sciences at the University of Virginia. Early in his career Singer designed the first satellite instruments to measure cosmic radiation and ozone.

In 1990, Fred Singer founded the Science and Environmental Policy Project (SEPP), which has received funding from Exxon, Shell, Unocal, ARCO, and according to SourceWatch, the Reverend Sun Myung Moon (who believes that he, Moon, is the Messiah and the second coming of Christ). In answer to the question of whether global warming would be good or bad, the SEPP website answers, "Probably both, but warming is definitely better than cooling."[2]

Singer has a record of denial on topics that include tobacco smoke, ozone depletion, acid rain, and toxic waste. In the early 1990s he cautioned energy companies that they could meet the same fate as the chemical companies after CFCs were discovered to deplete the ozone layer, which protects us from dangerous ultraviolet rays. "It took only five years to go from . . . mandating a simple freeze of production [of CFCs] at 1985 levels, to the 1992 decision of a complete production phase-out," he warned.[3]

In February 1992, Singer's SEPP published a "Statement by Atmospheric Scientists on Greenhouse Warming," intended as a preemptive strike against the outcome of the upcoming Rio Earth Summit, which according to Singer's statement:

> Aims to impose a system of global environmental regulations, including onerous taxes on energy fuels, on the population of the United States and other industrialized nations. Such policy initiatives derive from highly uncertain scientific theories. They are based on the unsupported assumption that catastrophic global warming follows from the burning of fossil fuels and requires immediate action. We do not agree.[4]

Among the signers were two other scientist-deniers profiled in this chapter, Richard Lindzen and Patrick Michaels.

SEPP spawned the International Center for a Scientific Ecology (ICSE), based in Paris and headed by science journalist Michel Salomon.

Singer and Salomon organized a conference in April 1992 in Heidelberg that produced the Heidelberg Appeal, designed to rebut the findings of the IPCC. Published at the close of the Rio Summit, the Heidelberg Appeal warned governments not to base their decisions on "pseudo-scientific arguments or false and non-relevant data. . . . The greatest evils which stalk our Earth," the appeal continued, "are ignorance and oppression, and not Science, Technology and Industry."[5]

One of the most useful resources to those interested in the role of industry in shaping public policy and denying global warming is Source-Watch, a collaborative project of the Center for Media and Democracy.[6] According to SourceWatch, the Heidelberg Appeal "was a scam perpetrated by the asbestos and tobacco industries in support of the Global Climate Coalition" (profiled in chapter 9).[7] Asbestos and tobacco companies had no direct quarrel with global warming, but wanted to promote their claim that there are two kinds of science: sound science (which favors our position) and junk science (which favors our opponent's position). The SEPP and ICSE (which also received funding from the vinyl and chemical industries), are typical denier organizations, appearing to be legitimate scientific think tanks while doing the dirty work of corporations.

In 1993, Philip Morris created The Advancement of Sound Science Coalition (TASSC). Singer collaborated for TASSC on an article about "junk science" and helped the Alexis de Tocqueville Institution in its effort to prevent regulation of secondhand smoke. Serving as self-appointed "director" of TASSC was Stephen J. Milloy, a career-long lobbyist for tobacco and other corporations that oppose government regulation of their products. Today Milloy runs a website called Junk-science.com.

Just as the global warming deniers now attack the IPCC, so TASSC attacked the Environmental Protection Agency and its finding that secondhand smoke was a health hazard. To hide its support for TASSC, Philip Morris chose a new PR firm called APCO Associates. "The credibility of the EPA is defeatable," said one Philip Morris internal document, "but not on the basis of environmental tobacco smoke alone. It must be part of a large mosaic that concentrates all of the EPA's enemies against it at one time."[8] To protect its profits, Philip Morris was ready to declare war on science and on environmental protection.

TASSC added Fred Singer to its board of advisers, along with Patrick Michaels and Frederick Seitz, whom I will profile later. It made plans with APCO to create a TASSC analog in Europe, listing "global warming" first among the new topics with which Big Tobacco's campaign of denial could align itself.[9]

TASSC shut down in 1998, but Milloy resurrected it as "The Advancement of Sound Science Center," which he ran out of his home. Between 2000 and 2004, ExxonMobil gave Milloy's center $50,000, while $60,000 went to "The Free Enterprise Education Institute," which also uses Milloy's home address.

In 1996, Singer's SEPP launched a publicity stunt promoting the "Top Five Environmental Myths of 1995":

(1) Global warming and the climate treaty;
(2) Stratospheric ozone and skin cancer;
(3) The secondhand smoke scare;
(4) The phony radon scare;
(5) The wasteful pursuit of zero risk, which mandates banning "the addition to food of any substance that has been found to cause cancer when fed in huge doses to rats."[10]

To help promote criticism of the EPA, Singer agreed to an "aggressive media interview schedule" organized by a PR firm working for British American Tobacco.

In 1993, under oath in a sworn affidavit, Singer admitted that oil companies funded his climate change research. Yet in 2001 he wrote to the *Washington Post* denying that in the last two decades he had received any oil company money.[11]

Another invaluable source of information about the deniers, ExxonSecrets, provides a list of Singer's "greatest quotes (ExxonSecrets is a project of Greenpeace):

"Climate science does not support the Kyoto protocol. . . . The climate is not warming and climate models have not been validated. In any case, a warmer climate would be generally beneficial."

"There is no convincing evidence that the global climate is actually warming."

"Even if the climate is warming, it does not mean it is due to human activities."

"To sequester carbon dioxide . . . implies carbon dioxide is bad when it's not bad, it's good. We should have more carbon dioxide in the atmosphere. It's good for plants. It makes them grow faster."

"The claimed consensus on substantial future warming simply does not exist. . . . The 20th century has turned out not to be 'unusual.' . . . The most accurate measurement of atmospheric temperatures from weather satellites showed little, if any, current warming."

"There is no proof at all that the current warming is caused by the rise of greenhouse gases from human activities."[12]

In March 2008, ABC News profiled Singer, leading to this exchange:

DAN HARRIS (ABC anchorman) : There are so many scientists who disagree with what you're saying, the IPCC [Intergovernmental Panel on Climate Change], NASA, NOAA [National Oceanographic and Atmospheric Administration], the National Academy of Sciences, the American Association for the Advancement of Science, the American Geophysical Union, the American Meteorological Society—we're talking about scientists all over the globe.

S. FRED SINGER: What can I say? They're wrong.

HARRIS: All of those people are wrong?

S. FRED SINGER: Yes.[13]

With All Due Respect, Dr. Michaels

In *The Heat Is On,* author Ross Gelbspan describes how the Western Fuels Association, a $400 million consortium of coal suppliers and utilities, outlined its strategy to combat global warming in a 1994 report: "[T]here has been a close to universal impulse in the trade association community here in Washington to concede the scientific premise of global warming . . . while arguing over policy prescriptions that would be the least disruptive to our economy."[14] The report continued, "We have disagreed, and do disagree, with this strategy. . . . When [the climate change] controversy first erupted . . . scientists were found who are skeptical about much of what seemed generally accepted about the potential for climate change." Among the "skeptical" scientists who came to the aid of the fossil fuel companies were S. Fred Singer and Patrick J. Michaels, whom ExxonSecrets calls "possibly the most prolific and widely-quoted" scientist to deny global warming.

Michaels has a PhD in ecological climatology from the University of Wisconsin at Madison and a long record of publication in peer-reviewed journals. He has written several books, including *Climate of Extremes: Global Warming Science They Don't Want You to Know* (2009); *Meltdown: The Predictable Distortion of Global Warming by Scientists, Politicians, and the Media* (2005); and *Shattered Consensus: The True State of Global Warming* (2005).

Michaels is editor in chief of the *World Climate Review*, a newsletter and blog funded by the Western Fuels Association, whose mission is to point out "the weaknesses and outright fallacies in the science that is being touted as 'proof' of disastrous warming." Gelbspan reports that over four years Michaels received more than $115,000 from oil and coal companies; ABC News said that the Intermountain Rural Electric Association "contributed $100,000 to Dr. Michaels."[15]

The libertarian Cato Institute, where Michaels is senior fellow, is not taxpayer supported, but subsists in part on donations from corporations like ExxonMobil, General Motors, and R. J. Reynolds, as well as right-wing institutions such as the Koch, Olin, and Scaife Foundations.

On February 12, 2009, Michaels testified before the House Energy and Commerce Subcommittee: "We often hear that 'the science is settled' on global warming. This is hardly the case. . . . Our [climate] models are in the process of failing. If it is demonstrable that these models have failed," he continued, "then there is no real scientific basis for any estimates of the costs of inaction."[16] But as I will show in chapter 12, the models have not failed.

In March 2009, on behalf of the Cato Institute, Michaels circulated an advertisement directed at President Barack Obama, who had said on November 18, 2008, "Few challenges facing America and the world are more urgent than combating climate change. The science is beyond dispute and the facts are clear."[17] Michaels's statement in response read:

With all due respect Mr. President, that is not true.

We, the undersigned scientists, maintain that the case for alarm regarding climate change is grossly overstated. Surface temperature changes over the past century have been episodic and modest and there has been no net global warming for over a decade now. After controlling for population growth and property values, there has been no increase in damages from severe weather-related events. The computer models forecasting rapid temperature change abjectly fail to explain recent climate behavior.

To back up these assertions, the ad included four citations to the scientific literature.

RealClimate.org analyzed both Michaels's House testimony and the ad, concluding, "[The ad] is a classic red-herring: ignore the facts you don't dispute, pick some others that are ambiguous and imply that, because they are subject to some debate, we therefore know nothing. Michaels (and Cato) presumably think this kind of nonsense is politically useful and he may be correct. But should he claim it is scientifically defensible, we would have to answer, 'With all due respect, Dr. Michaels, that is not true.'"[18]

The $45 Million Man

The most distinguished denier of global warming no doubt was Dr. Frederick Seitz, who died in March 2008 at age 96, a few days before a Heartland Institute Conference where he was to give a talk titled, "Predictions of Harmful Climate Effects Do Not Conform to the Evidence." In December 2001, Seitz and Robert Jastrow wrote an article in *Environment and Climate News* in which they said, "We find the scientific evidence clearly indicates the global warming of the last 100 years is likely not due mostly to human activities."[19] Jastrow had been a longtime proponent of the view that the Sun is the cause of any observed global warming.

Seitz held a bachelor's degree in physics from Stanford and a PhD from Princeton, earned in only two years. A solid-state physicist, he served as president of Rockefeller University and president of the

National Academy of Sciences. In 1984 Seitz joined with two other members of an obscure society of government advisers calling themselves the Jasons to establish the George C. Marshall Institute, today one of the most prominent organizations to deny global warming science.

In 2006, *Vanity Fair* magazine called Seitz the "$45 million man," referring to the sum he had helped the R. J. Reynolds tobacco company, whom he served as principal scientific adviser, distribute for its "research program," which actually was aimed at denying that smoking harmed health. In the article, Seitz admitted that, "They [Reynolds] didn't want us looking at the health effects of cigarette smoking." For aiding and abetting Reynolds in its campaign to create doubt about the dangers of smoking, Seitz earned $585,000.[20]

Seitz originally entered into a contract with Reynolds in 1976. But his services were available to other tobacco companies as well. One of the documents on the Tobacco Legacy website, the trove of more than 11 million documents that Big Tobacco was required to disgorge in response to the suit from state attorneys general, is a report from the manager of corporate scientific affairs for Philip Morris. He wrote that the company's strategy was to

> Publicize and communicate the results of a Marshall Institute report that challenges the scientific basis of various environmental regulations. The report was written by Dr. Frederick Seitz. In addition to his criticisms of the global warming and ozone depletion issues, Dr. Seitz also addresses the environmental tobacco smoke issue. Dr. Seitz concluded that "there is no good scientific evidence that moderate passive inhalation of tobacco smoke is truly dangerous under normal circumstances." The report will be used to challenge the EPA's report on environmental tobacco smoke in domestic and international markets."[21]

After doing the dirty work of Big Tobacco, Seitz moved on to oppose action on global warming. As told in chapter 2, in 1988 the Oregon Institute of Science and Medicine distributed a petition calling on scientists to show that they rejected global warming. Accompanying the petition was a letter of endorsement from Seitz, identifying himself as "Past President, National Academy of Sciences, U.S.A." and "President

Emeritus, Rockefeller University." Though signed only by Seitz, the letter used the phrase "in our opinion," which may have caused some readers to regard "our" as a reference to the Academy, the university, or both.

Also accompanying the petition was the paper in the format of the *Proceedings of the National Academy of Sciences* that Robinson had typeset on his home computer. In an interview, Robinson explained, "I used the *Proceedings* as a model, but only to put the information in a format that scientists like to read, not to fool people into thinking it is from a journal."[22]

The combination of the *PNAS*-like paper and the cover letter from Seitz was bound to give recipients the impression that they were receiving a peer-reviewed scientific paper so important that a past president of the National Academy of Sciences had given his imprimatur and helped to distribute it. The abstract of the paper read in part: "A review of the research literature concerning the environmental consequences of increased levels of atmospheric carbon dioxide leads to the conclusion that increases during the 20th and early 21st centuries have produced no deleterious effects upon Earth's weather and climate. Increased carbon dioxide has, however, markedly increased plant growth. Predictions of harmful climatic effects due to future increases in hydrocarbon use and minor greenhouse gases like CO_2 do not conform to current experimental knowledge."[23]

The Academy promptly disassociated itself from Seitz and the supposed research report that had accompanied his letter:

> The Council of the National Academy of Sciences (NAS) is concerned about the confusion caused by a petition being circulated via a letter from a former president of this Academy. The NAS Council would like to make it clear that this petition has nothing to do with the National Academy of Sciences and that the manuscript was not published in the *Proceedings of the National Academy of Sciences* or in any other peer-reviewed journal. The petition does not reflect the conclusions of expert reports of the Academy.[24]

On August 31, 1989, one year after Seitz wrote his letter accompanying the petition, Philip Morris decided it could no longer use him. Bill Murray of the company wrote that "Dr. Seitz is quite elderly and not sufficiently rational to offer advice."[25]

If Science Doesn't Have Integrity

If Frederick Seitz reached loftier heights than any other denier, Richard Lindzen, the Alfred P. Sloan Professor of Meteorology at MIT introduced in chapter 1, has the most relevant credentials in climate science. Holder of a PhD from Harvard, member of the National Academy of Sciences, Lindzen's curriculum vitae lists more than 230 publications. He served on the National Academy of Sciences panel that in 2001 examined the evidence for global warming and helped prepare the 1995 and 2001 IPCC reports.

Lindzen's opposition to global warming goes back for at least two decades. He signed the Heidelberg Appeal of 1992 and the Leipzig Declaration of 1995, whose operative paragraph said:

> It has become increasingly clear that . . . there does not exist today a general scientific consensus about the importance of greenhouse warming from rising levels of carbon dioxide. . . . Most climate specialists now agree that actual observations from both weather satellites and balloon-borne radiosondes show no current warming whatsoever—in direct contradiction to computer model results. [26]

Joining Lindzen and Singer as signatories were Michaels, William Nierenberg (one of the Jasons), Seitz, and twenty-five television meteorologists.

Lindzen, like Crichton, Singer, and other global warming deniers, argues that secondhand smoke is not a health hazard. In a 2001 interview with Lindzen in *Newsweek*, reporter Fred Guteri wrote, "He'll even expound on how weakly lung cancer is linked to cigarette smoking. He speaks in full, impeccably logical paragraphs, and he punctuates his measured cadences with thoughtful drags on a cigarette."[27]

In 1989, Eugene Mallove reported for *M.I.T. Tech Talk* on a speech that Lindzen had given to a colloquium at the institute. "I personally feel that the likelihood over the next century of greenhouse warming reaching magnitudes comparable to natural variability seems small," Lindzen said, adding that, "The entire [temperature] record would more likely be saying that the rise is 0.1 degree plus or minus 0.3 degree [°C]."[28] Lindzen seems to have been predicting that global temperatures might even fall. "It seems to me," Lindzen concluded, "that if science doesn't have integrity, it isn't of much use to people."

Over the last two decades, the evidence for global warming has grown steadily stronger. The IPCC has evolved from uncertainty in 1990 to 90 percent certainty in 2007. We know that since 1900, temperatures have risen about 0.8°C (1.4°F), far more than Lindzen allowed. Scientists are supposed to recognize and have their views affected by new evidence, yet over this period Lindzen and the other deniers profiled in this chapter have not budged in their opposition. Lindzen closed his presentation at the 2010 conference of the Heartland Institute with these words: "The failure to improve the case [for anthropogenic global warming] over 20 years makes the case even less plausible as does the evidence from Climategate and other instances of overt cheating."[29]

What has changed is the vehemence with which the deniers state their opposition. If one is unable to rebut the IPCC's conclusions on the facts, then as the certainty of those conclusions grows, deniers must resort to stronger and stronger language, to the point that speakers at the 2009 Heartland Institute Conference mentioned the IPCC with contempt and ridicule. Lindzen was among the most blunt, accusing the IPCC and scientists who did not agree with him of selling out for money, as quoted in chapter 1.

If there were ample evidence against global warming, one would think that the deniers would have discovered and published it. Why have they not done so? In the *Wall Street Journal* of April 12, 2006, Lindzen supplied the answer: "Scientists who dissent from the alarmism have seen their grant funds disappear, their work derided, and themselves libeled as industry stooges, scientific hacks or worse."[30] Former Astronaut Harrison Schmitt, whom we will meet next, concurs, saying that scientists who disagree with the alleged consensus on global warming "are being intimidated," and adding, "They've seen too many of their colleagues lose grant funding when they haven't gone along with the so-called political consensus that we're in a human-caused global warming."[31] Lindzen must be the exception: since 1975 he has received a total of $3,455,354 in research grants from the National Science Foundation alone.[32]

To confirm that other scientists who dispute global warming have had worthy research proposals denied because of politics, those scientists might wish to post on the Internet the proposals they have submitted, indicating which were funded and which not, and including the reviewers' comments.

Lindzen is unique among deniers in having postulated a specific negative feedback that could reduce global warming.[33] In 2001 he proposed that a warmer ocean would reduce cirrus cloud cover, allowing more infrared radiation to escape the earth, reducing the greenhouse effect and a creating the much-needed negative climate feedback.

In his March 2009 speech to the Heartland Institute Conference, Lindzen reiterated his claim: "Nature is, as any reasonable person might suppose, dominated by stabilizing negative feedbacks rather than destabilizing positive feedbacks." Lindzen was saying that negative feedbacks must exist to offset the known amplifying feedbacks. If he is right, then climate sensitivity must be less than the estimated 3°C rise for doubled CO_2.

In November 2009, Lindzen, together with postdoctoral student Y. S. Choi, published an article in *Geophysical Research Letters* that claimed to have found further evidence for the missing negative feedback.[34] The paper used data from a NASA satellite that measured the gain and loss of energy in the tropical atmosphere, leading the authors to assert that the climate sensitivity was indeed much less than climate scientists and their models had assumed.

In a response in the same journal, four climate scientists concluded that Lindzen and Choi's methods have "nothing to say about climate sensitivity."[35] Furthermore, "[Lindzen and Choi's] results do not stand up to independent testing." One of the authors of the rebuttal, Kevin Trenberth, told former *New York Times* reporter Andy Revkin that the Lindzen-Choi paper's flaws were sufficient to make it appear that the two "contrived the answer they got." Trenberth said that when he rectified the flaws in the original paper, he obtained a climate sensitivity of 4.1°C, even higher than the figure from climate models. In a follow-up to Revkin, Lindzen admitted that some of the criticisms were valid and said that he and Choi would address them in a new paper.[36]

Events have overtaken Lindzen. Data from paleoclimates and from more recent warming trends (summarized in the chart on p. 143) show that climate sensitivity is plenty high enough to make a difference.

Blue Marble

The denier who has had the most exciting career is Harrison "Jack" Schmitt, the last man to set foot on the moon. Schmitt grew up in the small New Mexico mining town of Silver City, in what had once been Apache country in the far southwestern corner of the state. In the 2000 census, Silver City had only 10,000 people. From this remote small town, Schmitt went on to earn a bachelor's degree from the California Institute of Technology and a PhD in geology from Harvard. He was in the first group of scientist-astronauts chosen by NASA and the only trained scientist ever to walk on the moon. Schmitt journeyed there in December 1972 on *Apollo 17* and may have taken the famous photograph called the "Blue Marble," looking back at Earth, a lonely sphere afloat in the cold abyss of space. In 1975 he resigned from the astronaut corps to run and win election as Republican senator from New Mexico. He lost his reelection effort to state Attorney General Geoff Bingaman by a substantial margin.

Schmitt not only left the astronaut corps; somewhere along the way he left mainstream science, especially where global warming is concerned. From 1994 to 1998, Jack Schmitt served as chairman and president of the Annapolis Center for Science-Based Public Policy, whose chairman emeritus he remains. The Annapolis Center not only argues against global warming, but promotes logging as a way to improve forests and questions whether mercury is a threat to health, whether air pollution contributes to asthma, and whether pesticide residue on food is dangerous.

Between 1998 and 2005, the Annapolis Center received $763,500 million from ExxonMobil. At its annual dinner in 2004, the center honored Sen. James Inhofe, the greatest denier among sitting politicians, for "his work in promoting science-based public policy." The next year, the honoree was Congressman Joe Barton (R-TX), another prominent denier.

In 2008, Schmitt resigned from the Planetary Society, largely over the organization's endorsement of manned missions direct to Mars, the former astronaut preferring a plan that used the moon as a staging station on the way to the Red Planet. But in his letter of resignation, Schmitt noted that the Society had also recommended that the United States accelerate research into global climate change through more and better Earth observations:

> As a geologist, I love Earth observations. But, it is ridiculous to tie this objective to a "consensus" that humans are causing global warming when human experience, geologic data and history, and current cooling can argue otherwise. "Consensus," as many have said, merely represents the absence of definitive science. You know as well as I, the "global warming scare" is being used as a political tool to increase

government control over American lives, incomes and decision making. It has no place in the Society's activities.[37]

In an interview for the *Santa Fe New Mexican*, Schmitt explained his resignation further, saying that "indisputable facts" show that global warming is natural; that the sea has been rising for thousands of years; that Antarctic ice is increasing; and that after retreating for decades, in 2005 Greenland's glaciers began to advance.[38]

These claims and more Schmitt reiterated in his talk at the March 2009 Heartland Institute Conference, reserving special contumely for the IPCC and its reports. He ended his talk with the assertion that attempts to cap carbon emissions are nothing more than an effort to effect "an enormous transfer of wealth and liberty from the people to the government," not the only time in the talk he had tied global warming to an assault on American liberty. Schmitt thus gives us another useful clue to what motivates some deniers. Their objection is less about science than what they see as the use of government regulation to threaten our constitutional right to life, liberty, and the pursuit of happiness. To prevent that, they are ready to abjure science itself.

On July 14, 2009, the *New York Times* commemorated the upcoming fortieth anniversary of the first landing of men on the moon with several articles, including one headlined "Vocal Minority Insists It Was All Smoke and Mirrors."[39] The article reported that some 6 percent of Americans believe the lunar landings were faked, nothing more than a hoax designed to instill national pride, like a World War II Hollywood film aimed at boosting public morale. To conclude the story, the reporter interviewed someone who ought to know: Harrison Schmitt. The former astronaut said, "If people decide they're going to deny the facts of history and the facts of science and technology, there's not much you can do with them. For most of them," he continued, "I just feel sorry that we failed in their education."

Nuke Vietnam?

Climate scientists should have been delighted that the March 29, 2009, issue of the Sunday *New York Times Magazine* would feature as its cover story an eminent physicist presenting his views on global warming. At last, some due recognition for mainstream science and from the *New*

York Times! How scientists' hearts must have sunk when they saw staring back at them from the front cover the wizened face of longtime denier and career-long contrarian Freeman Dyson.[40] "The global-warming heretic," the caption read in institutional-green ink, continuing, "How did Freeman Dyson—revered scientist, a liberal intellectual, problem-solver—wind up infuriating the environmentalists?" Anyone who saw only the cover would have had no doubt that scientists disagree about global warming: even Freeman Dyson is a denier!

Dyson has had a successful career in science despite lacking a PhD, having made a number of important discoveries. But the 85-year-old scientist has never accepted global warming. "Most of the evolution of life occurred on the planet substantially warmer than it is now," he says, "and substantially richer in carbon dioxide." As to the observed rise in sea level, not to worry, says Dyson, "why this is and what dangers it might portend cannot be predicted until we know much more about its causes." He has "studied the climate models," finding that "they do not begin to describe the real world that we live in. The real world is muddy and messy and full of things that we do not yet understand. It is much easier for a scientist to sit in an air-conditioned building and run computer models, than to put on winter clothes and measure what is really happening outside in the swamps and the clouds. That is why the climate model experts end up believing their own models."[41]

Dyson's early experience with climate models comes from his membership in the Jasons, the group of scientists—mainly physicists—who spent their summer vacations performing studies for the Defense Department. Critics of the Vietnam War blamed the Jasons for what they considered the group's collusion with the American military. To counter that, as one Jason said, they "wanted to work on things so that [we] could—I'll put it a little too bluntly—go back and tell [our] colleagues that [we] were really working on peaceful rather than war-related projects."[42] Climate studies fit the bill nicely, as they were so innocuous as not even to require a security clearance. That no member of the group knew anything about climate and modeling did not faze the Jasons: they were supremely confident of their ability to enjoy a few summer weeks in beautiful La Jolla, California, and come up with answers that had eluded lesser mortals. As one oceanographer who resigned from the Jasons put it, "I did resent this 'the ocean is a wonderful summer playground for smart physicists who can do it in their spare time'" (138).

Rather than relying on someone else's computer model of climate, the Jasons built their own. Dyson was a regular member of the climate group. The models were necessarily simple, whereas climate is the most complex problem humans have ever tried to model. The strategy of the Jasons, pressed for time to make the next breakthrough before the summer ended, was to simplify their models even more. As one put it, "We did a little model-making and we had fun, a lot of fun. It was a nice little physics project" (136). If Dyson's last brush with climate models was in the 1970s, no wonder he scoffs at the models and derides those who use them.

One of the war-related Jason projects on which Dyson worked was called, "Tactical Nuclear Weapons in Southeast Asia," following up a suggestion of a former chairman of the Joint Chiefs of Staff, who had said at a cocktail party, "Well, it might be a good idea to throw in a nuke once in a while just to keep the other side guessing" (93). Recognizing that thinly populated locations like the Mu Gia Pass, an entry to the Ho Chi Minh trail, made excellent targets, Dyson remembered, "[Another Jason] and I then decided this was something we really had to take seriously" (94).

One of the first things Dyson and the Jasons discovered was that every U.S. military base in Southeast Asia was within range of Soviet missiles. If the Soviets chose to respond to America's use of nuclear weapons with its own, "the U.S. fighting capability in Viet Nam would be essentially annihilated" (96–97). Author Ann Finkbeiner concludes, "In short, using tactical nuclear weapons in southeast Asia was a bad idea because our side made a better target than the other side" (97). Dyson concurred, saying, "It was mostly just a question of how vulnerable we were and how invulnerable the other side was." But what would have been the conclusion, Finkbeiner wondered, if Dyson and the Jasons had found that dropping atom bombs on Vietnam had been a perfectly good idea? "We probably would have then kept quiet and not said anything," he responded (97).

"There's no doubt that parts of the world are getting warmer, but the warming is not global," Dyson says. "I am not saying," he goes on, "the warming doesn't cause problems. Obviously it does. Obviously we should be trying to understand it. I am saying that the problems are being grossly exaggerated."[43]

Dyson reserves special scorn for NASA's James Hansen:

Hansen has turned his science into ideology. He's a very persuasive fellow and has the air of knowing everything. He has all the credentials. I have none. I don't have a PhD. He's published hundreds of papers on climate. I haven't. By the public standard he's qualified to talk and I'm not. But I do because I think I'm right. I think I have a broad view of the subject, which Hansen does not. I think it's true my career doesn't depend on it, whereas his does. I never claim to be an expert on climate. I think it's more a matter of judgment than knowledge.

Hansen, says Dyson, is "the person really responsible for this overestimate of global warming. He consistently exaggerates all the dangers." Dyson blames Hansen and Al Gore for relying too heavily on climate models and accuses them of "lousy science" and of "distracting public attention" from "more serious and more immediate dangers to the planet." Global warming might even be a boon, he says; it might be warding off the next Ice Age.

Dyson likes to play the hard-to-convince skeptic, disdaining consensus science in preference to his own ideas, the wackier the better. In the 1950s, Dyson endorsed the concept of powering spacecraft with nuclear explosions. One of the reasons he is not worried about global warming is that he considers it likely that, one day, carbon-eating trees will gobble carbon dioxide from the atmosphere and either bury it underground, or—here's an even better idea—convert it into liquid fuels and other useful chemicals. These yet-to-be-invented trees, Dyson claims, would reduce the amount of carbon dioxide in the atmosphere by half within 50 years. In any case, carbon dioxide, he says, is really a problem of land management. If we increase the amount of topsoil, the problem would take care of itself. Of course, first we would have to offset the billions of tons of topsoil that the earth loses each year to erosion.

"My first heresy says that all the fuss about global warming is grossly exaggerated," Dyson says. "Here I am opposing the holy brotherhood of climate model experts and the crowd of deluded citizens who believe the numbers predicted by computer models. Of course, they say, I have no degree in meteorology and I am therefore not qualified to speak. But I have studied the climate models and I know what they can do."[44]

In an interview on National Public Radio, Nicholas Dawidoff, who wrote the *New York Times Magazine* profile on Dyson, revealed how little thought he, Dawidoff, had given to the profile's effect on the public:

NICHOLAS DAWIDOFF: When people feel strongly about something and when it's a matter of great urgency and when it's a matter, for many people, of a looming apocalypse, of course, it should be taken very seriously. You definitely always want to hear from people who are going to push back against consensus. It only makes the people who are the majority or the people who are going forward and making public policy sharpen their arguments.

BOB GARFIELD (NPR): Does it matter, from a journalistic point of view, whether [Dyson is] right or whether he's wrong?

NICHOLAS DAWIDOFF: Oh, absolutely not. I don't care what he thinks. I have no investment in what he thinks. I'm just interested in how he thinks and the depth and the singularity of his point of view.[45]

Dawidoff's profile, featuring a global warming denier in one of the most prominent spots in American journalism, brought an outpouring of anger and denunciation. Blogger "Gail" may have spoken for them in a letter she addressed to the *New York Times*: "What is the matter with you? Do you realize we are on the verge of a global ecosystem collapse and runaway greenhouse warming? Don't any of you have children?"[46]

The *Times* profile certainly raised Dyson's standing. In its November 2009 issue, the *Atlantic* listed him as one of its 100 "Brave Thinkers."[47] Why? "He's taking a contrarian view on the Kyoto Protocol." The piece ends by lauding Dyson for "a lifetime of scientific rigor and intellectual honesty." Can one deny the entire field of climate science, based only on opinions, and be intellectually honest?

Foot Soldier

The deniers have their generals, some quite celebrated, but any army must have its GIs who do the actual fighting and who do not wind up on the cover of national magazines. An example is Dr. Timothy Ball, a Canadian and "renowned environmental consultant and former professor of climatology at the University of Winnipeg." In the last decade, Ball has given over 600 public talks on science and the environment. Between 2002 and 2007, he published thirty-nine opinion pieces and thirty-two letters to the editor in twenty-four Canadian newspapers.[48] He appeared in the documentary *The Great Global Warming Swindle* and in Glenn Beck's Fox News special, "Exposed: The Climate of Fear."

Ball was a member of the Calgary-based Friends of Science, which the *Toronto Globe and Mail* exposed as funded by oil and gas companies. He left Friends of Science to chair the Natural Resources Stewardship Project. Two of its three directors are executives of a public relations and lobbying company that works for energy industry clients. The project's first campaign "is focused on dispelling the notion that Canada needs CO_2 reduction plans," claiming that "CO_2 is very unlikely to be a substantial driver of climate change and is not a pollutant. Climate change is primarily a natural phenomenon." Ball also writes for Tech Central Station, an industry-funded denier website.[49]

In April 2006, Ball published an opinion piece on global warming in the *Calgary Herald*. The introduction described him as Canada's first PhD in climatology and a professor of that subject at the University of Winnipeg for twenty-eight years. In the article, Ball disparaged two other Canadian scientists. One of the two, Dan Johnson, Professor of Environmental Science at the University of Lethbridge, wrote to the *Herald* pointing out that Ball could not have been a professor for as long as claimed because he received his PhD only in 1983. Ball was later to admit that he had served for eight years, not twenty-eight, and that his degree is in geography, not climatology. The letter also alleged that during his professorial appointment, Ball did "not show any evidence of research regarding climate and atmosphere."

Ball sued the newspaper, its publishers and editors, the University of Lethbridge, and Professor Johnson, claiming that the letter had damaged his reputation as an environmental consultant and author, as well as his income-earning capacity "as a sought after speaker with respect to global warming."

Ball's suit backfired. In its response, the *Herald* expressed confidence in Johnson's letter and dismissed Ball's "credibility and credentials as an expert on the issue of global warming," adding, "The plaintiff is viewed as a paid promoter of the agenda of the oil and gas industry rather than as a practicing scientist." In June 2007, Ball withdrew the suit.[50]

Ball says global warming is "the political agenda of a group of people who believe that industrialization and development and capitalism and the Western way is a terrible system and they want to bring it down."[51]

Fear of State

The Nonscientists

On September 25, 2005, three-and-a-half weeks after Hurricane Katrina struck New Orleans, host George Stephanopoulos of ABC's *This Week* pondered with panelists Donna Brazile, David Gergen, and George Will whether global warming had affected hurricane intensity.[1] Gergen argued that the connection only made common sense: We know that waters in the Gulf of Mexico have warmed. Warmer waters should cause more severe hurricanes. Two new studies had just shown that Atlantic hurricanes had become more severe. Gergen detected cause and effect, but Will begged to differ. With a sigh of ennui, the *Washington Post* pundit protested:

> I have an alternative theory. I think these two hurricanes were caused by the prescription drug entitlement. You will say, "How can you say that? The entitlement hasn't even started." There's no conclusive evidence that global warming, that is to say, an unprecedented, irreversible, and radical change has started. You will say, "There's no scientific proof." Same answer. You will say, "Aren't you embarrassed, Mr. Will, to be attaching your political agenda to a national disaster?" Yeah, I'm embarrassed, but everyone else is doing it.

Aren't You Embarrassed, Mr. Will?

For George Will to have agreed with Gergen, he would have had to give up his by then fifteen-year long, now twenty-year long, derision of global warming and all who find it plausible.

Will holds a PhD from Princeton in philosophy and is a former college professor and Pulitzer Prize winner. No knee-jerk conservative, Will condemned the Nixon presidency, criticized George W. Bush for his invasion of Iraq, ridiculed the McCain-Palin presidential ticket, and scoffed at creationism and its latest incarnation, intelligent design. But on global warming, George Will has been the staunchest of deniers. He exemplifies the nonscientist-deniers profiled in this chapter.

In the spring of 2009, Will received more attention than he bargained for in response to a column titled, "Dark Green Doomsayers."[2] Though for two decades his position on global warming had remained unchanged, a sea change in communication had taken place: the blogosphere. No sooner had Will's column appeared than bloggers began to rebut each of its three principal claims.

For his first claim, Will repeated a favorite assertion: that in the 1970s scientists had predicted a coming ice age. In support, Will wrote: "In the 1970s, 'a major cooling of the planet' was 'widely considered inevitable' because it was 'well established' that the Northern Hemisphere's climate 'has been getting cooler since about 1950.'" The source of the quotation was neither a scientist nor a scientific journal, but the *New York Times* of May 21, 1975. In the same paragraph, to bolster the claim that scientists in the 1970s were worried about global cooling, Will strung together mix-and-match quotations from seven other sources. In order they are: *Science News, Science, Global Ecology, International Wildlife, Science Digest, Newsweek,* and the *Christian Science Monitor.* Five of the seven are not scientific journals, but popular press publications. *Science* magazine is of course a peer-reviewed journal, with *Nature* one of the two most prestigious. It turns out the article to which Will referred was one of the classic papers in science, one that solved the mystery that had puzzled Tyndall, Arrhenius, Callendar, and Earth scientists everywhere: the cause of the Ice Ages. The paper, by J. D. Hays, John Imbrie, and N. J. Shackleton and titled "Variations in the Earth's Orbit: Pacemaker of the Ice Ages," demonstrated that the "Milankovitch cycles" had caused the Ice Ages.[3] These cycles are tiny changes in the earth's orbit and axial tilt that cool some regions enough to start ice accumulating and advancing at one time, only to change in the other direction thousands of years later and start the ice melting and retreating. After presenting the evidence for the theory, the authors wrote that "the results indicate that the long-term trend over the next 20,000 years is toward extensive northern hemisphere glaciation and cooler climate."

So, yes, the *Science* paper did predict global cooling: 20,000 years in the future. Will not only got it wrong, he misquoted and misused one of the greatest papers in science.

What of the other possibly scientific source: *Global Ecology*? It turns out to be an edited book titled *Global Ecology: Readings Towards a Rational Strategy for Man*, published in 1971.[4] The quote comes from a chapter in the book by meteorologist Reid Bryson, taken from a 1968 article he wrote for a non-peer-reviewed magazine called *Weatherwise*. Bryson was not so much trying to forecast future temperatures as to explain why temperatures since World War II had been declining. Bryson wrote: "The continued rapid cooling of the earth since World War II is in accord with the increase in global air pollution associated with industrialization, mechanization, urbanization, and exploding population, added to a renewal of volcanic activity."

Thus in "Dark Green Doomsayers," Will does not quote a single *scientific* source to support his allegation of widespread belief in global cooling during the 1970s, instead forging a farrago of misleading snippets. Will might have cited Lindzen, who said of the 1970s "coming Ice Age" myth in a 1992 review for the Cato Institute, "The scientific community never took the issue to heart, governments ignored it, and with rising global temperatures in the late 1970s the issue more or less died."[5] Or he could have used a 2008 review in the *Bulletin of the World Meteorological Organization*, which reported that between 1965 and 1979, forty-two peer-reviewed papers indicated global warming, while seven favored global cooling. The reviewers summed up: "There was no scientific consensus in the 1970s that the earth was headed into an imminent ice age. Indeed, the possibility of anthropogenic warming dominated the peer-reviewed literature even then."[6]

It is true that during the 1970s global temperatures remained flat, or even declined slightly. Suppose that toward the end of that decade, before climate models and understanding had advanced to anything like their present state, scientists *had* said that the earth could be entering a new cooling period, even approaching an ice age. They would have made an honest mistake, based on their interpretation of the evidence, a mistake that new findings and subsequent work by themselves and other scientists would have corrected. Their error would not mean that today, thirty or forty years later, armed with a mountain of new evidence and supercomputers, scientists must be wrong again. If we applied the "one-mistake-and-you're-out" rule to pundits, the op-ed pages would be empty.

For his second claim in "Dark Green Doomsayers," Will said, "As global levels of sea ice declined last year, many experts said this was evidence of man-made global warming. Since September, however, the increase in sea ice has been the fastest change, either up or down, since 1979, when satellite record-keeping began. According to the University of Illinois Arctic Climate Research Center, global sea ice levels now equal those of 1979."

Scientists from the Center responded: "We do not know where George Will is getting his information, but our data shows [sic] that on February 15, 1979, global sea ice area was 16.79 million sq. km and on February 15, 2009, global sea ice area was 15.45 million sq. km. Therefore, global sea ice levels are 1.34 million sq. km less in February 2009 than in February 1979. This decrease in sea ice area is roughly equal to the area of Texas, California, and Oklahoma combined."[7]

That left only Will's third argument: "According to the World Meteorological Organization, there has been no record of global warming for more than a decade, or one third of the span since the global cooling scare." But 1998, the year Will chose to start his comparison, had seen an unusually strong El Niño, the periodic tropical disturbances that lead to higher-than-average land surface temperatures.[8] Had Will chosen 1997 as his start date, he would have had to come to exactly the opposite conclusion.

How did George Will respond to the criticism? In a column on February 27, titled "Climate Science in a Tornado," he wrote, "The [February 15] column contained many factual assertions but only one has been challenged. The challenge is mistaken."[9]

When the brouhaha refused to die down, the *Post*'s ombudsman entered the fray with a column on March 1. He began by saying that "opinion columnists are free to choose whatever facts bolster their arguments. But they aren't free to distort them." The article went on to describe the process that the *Post* uses for fact-checking. It focused on Will's claim about global sea ice while ignoring the other two errors, thus seeming to accept Will's assertion that scientists had questioned only one of his original claims, when in fact they had questioned all three.[10]

On March 21, evidently realizing that it had dug an even deeper hole for itself, the *Post* took the unusual step of publishing two debunkings of Will's columns, one by science writer Chris Mooney and the other by the secretary-general of the World Meteorological Organization. Mooney's response included some important words: "Perhaps the only hope involves taking a stand for a breed of journalism and commentary . . . constrained by standards of evidence, rigor and reproducibility that are similar to the canons of modern science itself."[11]

In his letter, the secretary-general of the WMO, Michael Jarraud, said:

> It is a misinterpretation of the data and of scientific knowledge to point to one year as the warmest on record—as was done in a recent *Post* column . . . and then to extrapolate that cooler subsequent years invalidate the reality of global warming and its effects.
>
> The difference between climate variability and climate change is critical, not just for scientists or those engaging in policy debates about warming. Just as one cold snap does not change the global warming trend, one heat wave does not reinforce it. Since the beginning of the 20th century, the global average surface temperature has risen 1.33 degrees Fahrenheit.
>
> Evidence of global warming has been documented in widespread decreases in snow cover, sea ice and glaciers. The 11 warmest years on record occurred in the past 13 years.[12]

On April 2, Will returned to global warming in a third column, this one titled "Climate Change's Dim Bulbs," ridiculing the use of compact fluorescent light bulbs: "Warming is allegedly occurring even though, according to statistics published by the World Meteorological Organization, there has not been a warmer year on record than 1998."[13]

Having long ago given up on George Will, scientist-bloggers began to go after the *Washington Post* for its failure to fact-check Will's claims. The *Post* responded that a veritable army of fact-checkers review Will's columns: his own employees, editors at the *Post* writer's group, its op-ed page editor, and two copy editors.

The next step in the saga was the publication of a news article in the *Post* on April 7: "New Data Show Rapid Arctic Ice Decline: Proportion of Thicker, More Persistent Winter Cover Is the Lowest on Record." The article said that, "The new evidence—including satellite data showing that the average multiyear wintertime sea ice cover in the Arctic in 2005 and 2006 was nine feet thick, a significant decline from the 1980s—contradicts data cited in widely circulated reports by *Washington Post* columnist George F. Will that sea ice in the Arctic has not significantly declined since 1979."[14]

Finally, on April 11 the *Post* published an editorial titled, "Arctic Ice Is Melting: The 30-Year Decline Is Accelerating, New Data Show." The piece began, "Make no mistake, Arctic Sea ice is melting."

Once Again

Michael Crichton's techno-thriller *State of Fear* had an initial print run of 1.5 million copies and soon took the number one sales position on Amazon.com and the number two slot on the *New York Times* bestseller list. By the time *State of Fear* appeared in 2004, Crichton had built a loyal audience ready to turn any book he wrote into a bestseller and make the multimillionaire author and television screenwriter even wealthier.

Crichton, who died in November 2008 of lung cancer, was a tall, movie-star handsome graduate of Harvard Medical School. His first novel, *The Andromeda Strain*, did so well that he left medicine for a career in writing. *State of Fear* broke new ground for Crichton, as he attempted to write three books in one: (a) a science fiction novel, (b) a scientific report on global warming, and (c) an anti-environmentalist screed.

In the book, conscienceless environmental eco-terrorists, part of an organization called the Environmental Liberation Front, attempt to fake natural-seeming disasters to persuade the public of the dangers of global warming. Explosives would blast icebergs off Antarctica; rockets would cause lightning storms in national parks; a man-made tsunami would drown the California coast. An estimable MIT professor, whom some say Crichton modeled after himself, thwarts the eco-terrorists and saves the world.

State of Fear contains dozens of footnotes, many charts, and a 20-page bibliography, giving the impression that the book is so well-researched that Crichton grounded its fictional plot in scientific truth. According to *Publishers Weekly*, Crichton spent three years researching the book.

A reader who takes Crichton at his word would come away believing that

• The science behind global warming is weak and disputed by scientists themselves. We know far too little to try to curtail greenhouse gas emissions.

- Scientists are perfectly willing to distort their findings in order to achieve their liberal, or fascist, political goals. Politicians and others are prepared to join scientists in politicizing science.
- Gullible media, intelligentsia, and members of the general public are all too ready to adopt the latest doomsday fad, allowing deceptive scientists to dupe them.

The telegenic Crichton and his publisher promoted *State of Fear* widely, to praise and criticism. The *Wall Street Journal* loved the book, correctly describing it as a novelization of a speech the author gave at the Commonwealth Club of San Francisco in 2003. In that speech Crichton had said, "Environmentalism is essentially a religion, a belief-system based on faith, not fact."[15] According to the *Journal*, "To make this point, [*State of Fear*] weaves real scientific data and all too real political machinations into the twists and turns of its gripping story."[16]

As usual, the *New York Times* did not see things the same way as the *Journal*. The *Times* reviewer wrote, "Although *State of Fear* comes dressed as an airport-bookstore thriller, Michael Crichton readers will discover halfway through their flight that the novel more closely resembles one of those . . . 'Liberals are Stupid' jobs. To claim that [global warming] is a hoax is every novelist's right. To criticize the assumptions and research gaps in global warming theory is any scientist's prerogative. Citing real studies to support the idea of a hoax is ludicrous."[17]

The Pew Center on Global Climate Change summed up: "Although Crichton attempts to use real-world data and studies within the novel to highlight some of the realities and uncertainties in climate science, the novel contains a number of straw man arguments, misinterpretations of the scientific literature, and even a few misleading statements drawn from the so-called 'skeptics.'" The review continues: "Despite his research and the book's many footnotes, Crichton has a less-than-commanding understanding of climate change science. The book is much more of a vehicle for his own opinions on the issue rather than an objective commentary on the state of the science and policy debate."[18]

Scientists, science journalists, and science organizations generally panned *State of Fear*. James Hansen said, "Crichton writes fiction and seems to make up things as he goes along. He doesn't seem to have the foggiest notion about the science that he writes about."[19] Said a physics professor at New York University, "Wrong, wrong, wrong." He went on: "The best face I can put on this is that he doesn't know what he's doing. The worst is that he's intentionally deceiving people as he accuses environmentalists [of doing]."[20]

Richard Lindzen was an exception among scientists, saying *State of Fear* "was a fun read and the science was handled intelligently and responsibly. Crichton has studied the science for the last three years and comes to the issue with intelligence as well as a professional scientific background."[21]

In 2006 the American Association of Petroleum Geologists gave Crichton its Journalism Award, saying that *State of Fear* "is fiction, but it has the absolute ring of truth." The ensuing controversy about giving a book of fiction an award for journalism caused the AAPG to rename the prize the "Geosciences in the Media" award.

State of Fear made a number of scientific claims, each of which scientists had already discredited. Consider the following sample and go to RealClimate.org for a complete rebuttal.

One character in the book asks how, since the earth cooled from 1940 to 1970 even though carbon dioxide levels in the atmosphere were increasing, scientists can be sure of the reason for the current warming. Crichton's three years of research should have taught him that global temperatures reflect not only carbon dioxide levels but also the amount of dust, sulfates, and ozone in the air, as well solar activity. As the IPCC and others pointed out before *State of Fear* was published, when climate models incorporate both human-caused and natural factors, the models account for the short-term ups and downs as well as the long-term rise of twentieth-century temperatures.

Another Crichton claim was that global temperature measurements are anomalously high because of the urban "heat island" effect—the higher temperature of cities as compared with the surrounding countryside. But in 2001, at about the time Crichton is said to have begun his research, the IPCC estimated that only about 0.05°C of the twentieth-century temperature increase was due to the higher temperatures of urban areas.

Crichton claimed that in Hansen's testimony to Congress in 1988, he overestimated subsequent global warming by 300 percent. In a paper that came out just after his testimony, not knowing how carbon dioxide emissions would increase, Hansen had presented three model simulations: one with exponentially increasing carbon dioxide, one with business-as-usual emissions, and the third with no increase in emissions after the year 2000. (That last one we can forget.) In his testimony,

Hansen used only the second scenario, which turns out to have done a remarkably good job of forecasting the actual temperature increase since 1988. The source of the persistent claim that Hansen erred by "300 percent" was the later testimony of Patrick J. Michaels, who used only the curve of exponential increase to bolster his claim that since Hansen erred, climate models cannot be trusted. Evidently Crichton's research led him to Michaels.[22]

The title of the appendix to *State of Fear* is "Why Political Science Is Dangerous." In it, Crichton claims that scientists have politicized global warming science, while the deniers have done nothing more than attempt to set the record straight. Yet James Hansen did not spend an hour in the Oval Office with the President—that was Crichton. Instead, the Bush administration tried to squelch Hansen. Sen. James Inhofe did not make Hansen's papers required reading for his Senate Committee on Environment and Public Works and invite Hansen to testify—those were *State of Fear* and Crichton.

Crichton's appendix begins with the discredited theory of eugenics, which reached its evil culmination in Nazi Germany's extermination of Jews and other people it deemed inferior. One wonders where Crichton is heading—why bring up the Nazis in a book about global warming?

From eugenics Crichton turns to Soviet agriculture under Joseph Stalin and Soviet biologist Trofim Lysenko, and his intent becomes clear. In both Nazi eugenics and Stalinist agriculture, governments subverted science to politics, costing many lives. Crichton writes disingenuously, "I am not arguing that global warming is the same as eugenics. But the similarities are not superficial."[23] No matter how he tries to disguise the fact, Crichton *is* comparing global warming to two of the great evils of the twentieth century.

After ending the section on Lysenkoism, Crichton opens a new one with these Lincolnesque words: "Now we are engaged in a great new theory that once again has strong support of politicians, scientists, and celebrities around the world. Once again, critics are few and harshly dealt with," an assertion that has never been backed up with evidence.

One paragraph in the appendix is worth parsing carefully:

"Once again," Crichton writes, "the measures being urged have little basis in fact or science." But it is the deniers' position that has no basis in fact or science.

"Once again, groups with other agendas are hiding behind a movement that appears high minded." True, that is exactly what the *deniers* are doing: using pseudoscience to foster their antiregulatory and libertarian agendas. It is the deniers, many of them funded by the fossil fuel industry, who claim that their efforts to prevent or delay action on global warming are prompted by only the highest principles: liberty, democracy, capitalism, and the American way of life. Scientists are not hiding behind anything. They are doing their research and publishing the results.

"Once again, claims of moral superiority are used to justify extreme actions." Both sides may assert moral superiority, but only one can be right. If the deniers are wrong and something like worst-case global warming ensues, people will die. The deniers are willing to take that risk, or fail even to recognize that it is a risk, in spite of what scientists tell us. Which is the morally superior position?

"Once again, the fact that some people are hurt and shrugged off because an abstract cause is said to be greater than any human consequences." Global warming is not an abstract cause. It is a reality. As this book is being written, a report from Columbia University, the United Nations University, and CARE International says that climate change will displace tens of millions of people in coming years, posing social, political, and security problems of an unprecedented dimension. "Within the next few decades, the consequences of climate change for human security efforts could be devastating."[24] During 2010, floods devastated Pakistan and Moscow suffered record heat and smoke. I do not claim that these disasters are concrete evidence of global warming: but they do show what it will be like. Who has the moral high ground, the scientists who warn of the danger, or the deniers like Crichton who say there is no danger?

"Once again, vague terms like sustainability and generational justice—terms that have no agreed definition—are employed in the service of a new crisis." To the contrary, those terms have well-understood definitions. According to the *Oxford English Dictionary*, sustainability means "able to be maintained at a certain rate or level: sustainable economic growth." In ecology, the term means "conserving an ecological balance by avoiding depletion of natural resources." The *Encyclopedia of Global Change* says that, "The central question posed by [intergenerational equity] is what, if any, are the current generation's obligations to future generations." Colleges use the concept to decide how much of their endowment earnings to spend on the current generation of students and how much to reinvest so that future generations will have the same benefits. Any educated person knows or can figure out what these terms mean.

The phrase "once again" gets drummed into one's head until one loses sight of to what it actually refers: that global warming is but another example of politicized science, following in the footsteps of Nazism and Lysenkoism.

The Lomborg Deception

In 2001 a book called *The Skeptical Environmentalist* came to opposite conclusions from those of the IPCC's just-released Third Assessment Report.[25] The book contained an astounding number of references, nearly 3,000 in all. Its main argument was that humanity faces many larger and more pressing environmental problems than global warming. Population growth, AIDS, hunger, pollution, water shortages, deforestation, and species loss are all immediate, and most of them, Lomborg claims, stem from poverty. If world leaders aim to do the most good for humanity in the shortest time, he says, rather than throwing money at a problem of the distant future that may turn out not even to exist, they should instead declare war on poverty.

The Skeptical Environmentalist became an instant bestseller. The *Economist* got behind the book even before it came out, publishing an advance essay by Lomborg that primed the sales pump. The magazine's subsequent review said, "This is one of the most valuable books on public policy—not merely environmental policy—to have been written for the intelligent general reader in the past 10 years. . . . *The Skeptical Environmentalist* is a triumph."[26] *New Scientist* labeled Lomborg a "welcome heretic."[27] The *Wall Street Journal* found the book "superbly documented and readable." Concluded the *Washington Post*, "[Lomborg's] richly informative, lucid book is now the place from which environmental policy decisions must be argued."[28]

Perhaps the most remarkable thing about the book's favorable reception was that reviewers and the press immediately elevated the opinion of a young, unknown nonscientist onto an equal or even a superior footing with the conclusions of trained scientists. The auxiliary message of *The Skeptical Environmentalist*, like that of *State of Fear*, seemed to be that any reasonably intelligent but untrained person willing to put in a year or two of work could show that a host of scientists were wrong.

Scientists and scientific journals soon weighed in. *Scientific American* questioned Lomborg's conclusion, which he had claimed surprised even himself, that "contrary to the gloomy predictions of degradation . . . everything is getting better," as well as the author's accusation that "a pessimistic and dishonest cabal of environmental groups, institutions and the media" is responsible for the unwarranted gloom. The magazine summed up:

> It is hard not to be struck by Lomborg's presumption that he has seen into the heart of the science more faithfully than have investigators who have devoted their lives to it; it is equally curious that he finds the same contrarian good news lurking in every diverse area of environmental science.[29]

A review in *Nature* magazine went further, saying that the book "reads like a compilation of term papers from one of those classes from hell where one has to fail all the students. It is a mass of poorly digested material, deeply flawed in its selection of examples and analysis."[30] The Union of Concerned Scientists wrote that "Lomborg's book is seriously flawed and fails to meet basic standards of credible scientific analysis. He uncritically and selectively cites literature—often not peer-reviewed—that supports his assertions, while ignoring or misinterpreting scientific evidence that does not. This consistently flawed use of scientific data is . . . unexpected and disturbing in a statistician."[31]

The correspondent for the BBC, in an online article titled, "Bjorn Lomborg's Wonderful World," admitted to being neither a statistician nor a scientist, but echoed the criticisms of *Scientific American*: "I am worried that on virtually every topic he touches, he reaches conclusions radically different from almost everybody else. That seems to suggest that most scientists are wrong, short-sighted, naïve, interested only in securing research funds, or deliberately dancing to the campaigner's tune. Most I know are honest, intelligent and competent. So it beggars belief to suppose that Professor Lomborg is the only one in step, every single time."[32]

The criticism of Lomborg reached a new and more serious level when a group of Danish scientists formally accused him of deliberately using misleading data and reaching flawed conclusions. They took their complaint to the Danish Committee on Scientific Dishonesty, which found that Lomborg fabricated data, discarded unwanted results, deliberately used misleading statistics, distorted his interpretation of conclusions, plagiarized, and deliberately misinterpreted others' results. Lomborg complained to the Danish Ministry of Science, Technology, and Innovation, which annulled the findings of the committee, largely because it had made procedural errors.

This prompted one of the original complainants to take the accusations against Lomborg into his own hands. Danish biologist Kare Fog created a website (www.lomborg-errors.dk) to catalog Lomborg's mistakes.[33] Fog reviews *The Skeptical Environmentalist* page by page and comes up with 110 actual errors and 208 flaws, for a total of 318, or about one mistake per page. Fog performs the same analysis on *An Inconvenient Truth*, finding that Gore's book and film contain a total of two errors and twelve flaws, for a total of fourteen mistakes.

In 2007, Lomborg published *Cool It*, a rehash of the arguments in *The Skeptical Environmentalist*, except that now he agreed that "global warming is real and manmade."[34] But as in the earlier book, Lomborg claimed we need not worry about it: "The costs and benefits of the proposed measures against global warming . . . is the worst way to spend our money. Climate change is a 100 year problem—we should not try to fix it in 10 years."

Lomborg's argument thus condenses to two points: global warming is not urgent; trying to prevent it will cost a lot of money while doing little good[35]

To bolster his case, in an op-ed piece in the *Guardian* Lomborg cites the 2007 IPCC report, which predicted a rise in sea level from from 0.18 to 0.59 meters.[36] Indeed, if scientists today believed that it is possible that sea level will rise no more than 0.18 meter—about 8 inches—then we could all relax. But Lomborg conveniently leaves out the IPCC's caveat: it forecasted a rise of 18–59 cm *plus an unknown extra rise* from ice sheet melting.[37] Scientists today regard 1 meter (3 feet) of sea level rise as likely and 2 meters as a possibility. Because of the long lead time, if we want to prevent such rises we have to start acting now, not wait until it becomes urgent, which is usually a poor way to avoid risk.

As to the cost of prevention, Lomborg estimates that cutting carbon dioxide emissions will cost about $20 per ton of CO_2 saved while doing only about $2 worth of good. The massive Stern report of the British Government comes to quite a different conclusion.[38] After an exhaustive analysis that considered all possible effects of global warming including costs to the environment and human health, Stern concluded that business-as-usual climate change would reduce global consumption by an average of about 20 percent. In other words, the average human being would have to get by on about one-fifth less of everything, including food. Those in the developed world could certainly manage that, Stern argued, though they would not like it, but what about the billion or so people who now get by on $1 a day or less? They have nothing to fall back on and many would starve. And they have done nothing to cause global warming. Contrast that figure with Stern's estimate of the cost of stabilizing carbon dioxide: 1 percent of GDP, or 1 percent of consumption, one-twentieth the cost of doing nothing. Later Stern raised his estimate to 2 percent of GDP, to reflect the accelerated pace of global warming. Economist Stern and his economic team came to exactly the opposite conclusion about the cost effects of global warming from political scientist and statistician Lomborg. Moreover, weaning society off fossil fuels and onto renewable energy will have many benefits beyond avoiding global warming. Cleaner air and water, independence from Middle Eastern petro-nations, less environmental damage, fewer oil spills, innovations and spillover from new industries, and so on.

In an interview in the *Sunday Times* of London in August 2009, Lomborg remained dismissive about the harmful effects of global warming.[39]

> A pending fuel crisis? Hysteria, he said. World hunger? Baloney: food was increasing. Species extinction? Rubbish. Disappearing forests? Tosh: forest cover had increased. Indeed, he proclaimed, nearly every indicator demonstrated that man's lot had vastly improved. "The world in decline is a litany we have heard so often that another repetition is almost reassuring," he said. "There is just one problem: it does not seem to be backed up by the available evidence."

Just in case global warming does turn out to be dangerous, Lomborg, ever the contrarian, has the remedy: a scheme to generate clouds that would reflect the Sun's energy back into space. A fleet of 1,900 wind-powered ships would ply the oceans, sucking up sea water and spewing it into the air, where it would condense and create the much-needed clouds.[40] Such large-scale experiments to offset global warming caused by CO_2 emissions we call "geoengineering." No one knows whether they will work or whether they may not do more harm than good. The notion seems to be that before we do anything to lower carbon emissions, we must develop and test water-spraying ships, giant mirrors that float in space and reflect sunlight, iron dumped in the ocean to fertilize carbon dioxide–eating plankton, and so on. If they do not work, or if they make global warming worse, well, we tried.

Were he still alive, C. Northcote Parkinson would warn us that if through various geoengineering schemes we do succeed in reducing the amount of global warming caused by a specified amount of atmospheric CO_2, we will just use the opportunity to burn additional fossil fuel. When we give up on geoengineering, as eventually we must, the atmosphere will retain the extra CO_2 and global temperature will rise higher than if we had never tried to engineer the climate of our planet.

Let us return to the question with which I begin this section on Lomborg: Could anyone, even a climate scientist, really have read and assimilated nearly 3,000 references? Here is the place to acknowledge that authors, this one included, can be notoriously thin-skinned. If a review is 99 percent positive, some authors, this one included, will obsess over the 1 percent that is negative. Imagine then an author having to read not merely a review that was 100 percent negative, but a book of 272 pages devoted to refuting every word the author wrote. Such is Howard Friel's *The Lomborg Deception*, published by Yale University Press in March 2010.[41]

Friel's aim is "to show that Lomborg's Theorem [his claim that anthropogenic global warming is "no catastrophe" and we need spend no money to prevent it] is grounded in highly questionable data and analysis, and [has] little if any factual or analytic basis." In a long chapter, Friel takes as a case study Lomborg's statement in *Cool It* that "once you look closely at the supporting data, the narrative [that polar bears are threatened] falls apart." Friel proceeds to track down every polar bear reference that Lomborg cites, showing that the author cherry-picked articles that supported his position while ignoring those that did not. Lomborg's claim that the polar bears are fine vanishes like a melting iceberg.

In his next chapter, Friel pursues the references that Lomborg cites in the first chapter of *The Skeptical Environmentalist*, titled "Things Are Getting Better," about 250 citations in all. Friel decides to go after only the first 29, relentlessly pursuing Lomborg's every word with the obsession of Lieutenant Phillip Gerard after fugitive Dr. Richard Kimble. To accomplish this task takes Friel 9,000 words, each one of which Lomborg no doubt read, since he posted a lengthy rebuttal on his website.

Friel sums up: "To similarly review all of the 2,930 endnotes in Lomborg's book would require a 900,000-word book comprising more than 100 chapters the size of this one." Fortunately for his readers, Friel knew when to quit.

Newsweek columnist Sharon Begley took it upon herself to check Friel checking Lomborg.[42] She concluded, "Although Friel engages in some bothersome overkill, overall his analysis is compelling." She picked three of Lomborg's contentions, which she said "Friel pretty much blows out of the water."

Lomborg's latest gambit is a film titled *Cool It*, whose website declares it to be "a bold new vision from Bjorn Lomborg."[43] According to Box Office Mojo, in its third weekend, which followed the Thanksgiving holiday, the film grossed $918.

Bed-Wetting, Messianic, Moaning-Minnies

"Where are they all today, those bed-wetting, messianic, moaning-minnies of the apocalyptic traffic light tendencies, those greens too yellow to admit they're red?" boomed the Viscount Monckton of Brenchley at the 2009 Heartland Institute Conference.[44] The American public first met Monckton in October 2007, when the *New York Times* and the *Washington Post* ran advertisements in which he challenged Al Gore to a duel, not with shield and short-sword, but with words. The question to face the pair, said Monckton, was: "That our effect on climate is not dangerous." He called on the former vice president to "step up to the plate and defend his advocacy of policies that could do grave harm to the welfare of the world's poor." The Heartland Institute says it spent $1.2 million promoting the debate, but to no avail, since Gore declined.

But who is the Viscount Monckton of Brenchley and what credentials has he that would allow him to provide a serious, science-based challenge to Gore? Monckton is a graduate in classics of Cambridge University who later obtained a diploma in journalism and served as adviser to Margaret Thatcher. According to a biographical sketch that accompanies a 23-page letter he wrote to presidential candidate John McCain, Monckton provided Prime Minister Thatcher with advice on "Warship hydrodynamics . . . psephological modeling . . . embryological research . . . hydrogeology . . . public-service investment analysis . . . public welfare modeling . . . and epidemiological analysis."[45]

While recuperating from illness several years ago, Monckton amused himself by looking into global warming, which was becoming a well-publicized issue. Monckton, who says he can do "radiative heat transfer calculations . . . standing on my head," concluded from back-of-the-envelope scribbling that "there is very little for us to worry about at all," thus "outsmarting at a stroke thousands of the world's scientists," as an interviewer put it.[46]

Monckton sent his conclusions about global warming to a friend, who passed them on to the editor of the *Sunday Telegraph*, which published them under the heading "Climate chaos? Don't believe it." The article received so many online hits that it crashed the *Telegraph*'s website. Eventually Monckton's *Telegraph* articles numbered 52 pages, prompting *Telegraph* columnist and author George Monbiot to write that they were a "mixture of cherry-picking, downright misrepresentation, and pseudoscientific gibberish. There is scarcely a line . . . which is not wildly wrong."[47] In a characteristic response, Monckton threatened to sue Monbiot. Nevertheless, the *Telegraph* articles launched Monckton into a new career as an expert on global warming.

After Al Gore declined to debate Monckton, the Science and Public Policy Institute, which lists Monckton as "Chief Policy Advisor," funded a film entitled *Apocalypse? No!* in which Monckton charges Gore and the IPCC with systematically falsifying and exaggerating the evidence for global warming. The film is available in DVD format on the SPPI website.[48]

Monckton has a habit of making provocative statements on a variety of subjects. In the January 1987 issue of the *American Spectator*, he wrote, "There is only one way to stop AIDS. That is to screen the entire population regularly and to quarantine all carriers of the disease for life. Every member of the population should be blood tested every month. . . . All those found to be infected with the virus, even if only as carriers, should be isolated compulsorily, immediately, and permanently."[49]

Sen. James Inhofe lists Monckton among the 400 prominent scientists who dispute that global warming is man-made. Rep. Joe Barton also invited Monckton to speak before his committee, saying, "I especially want to thank Lord Monckton for testifying. He is generally acknowledged as one of the most knowledgeable, if not the most knowledgeable, expert from a skeptical point of view on this issue of climate change."[50]

In a letter to Senators Jay Rockefeller (D-WV) and Olympia Snowe (R-ME) defending the right of free speech of the CEO of Exxon-

Mobil and calling on the two senators to "change or resign," Monckton described himself as a "Member of the Upper House of the United Kingdom legislature."[51] But the House of Lords says that "Christopher Monckton is not and has never been a Member of the House of Lords."[52]

The entry for Monckton in the Inhofe "Who's Who" faults the United Nations for not apologizing to him for having abolished the Medieval Warm Period. In retaliation, Monckton recommended that the UN abolish the IPCC. Monbiot explains that Monckton's claim that the IPCC had deep-sixed the Medieval Warm Period rested on two graphs comparing the IPCC's rendering of the temperature record over the last 1,000 years with actual measured and estimated temperatures over the same period. The trouble, according to Monbiot, is that the two graphs "are measuring two different things: global temperatures (the [IPCC's] progression) and European temperatures (Monckton's line) . . . [and] the scales are different."[53]

In his encyclopedic letter to presidential candidate McCain in 2008, titled "More in Sorrow Than in Anger," Monckton said that his climate work had "earned him the status of Nobel Peace Laureate. His Nobel Prize pin, made of gold recovered from a physics experiment, was presented to him by the Emeritus Professor of Physics at the University of Rochester."[54]

Monckton took McCain to task for visiting a wind farm and devoting an entire campaign speech to, as Monckton described the speech, "the apocalyptic vision of catastrophic anthropogenic climate change— a lurid and fanciful account of imagined future events that was always baseless, was briefly exciting among the less thoughtful species of news commentators and politicians, but is now scientifically discredited."

In an article written for his Science and Public Policy Institute, Monckton branded climate scientists as war criminals:

These evil pseudo-scientists, through the falsity of their statistical manipulations, have already killed far more people through starvation than "global warming" will ever kill. They should now be indicted and should stand trial alongside Radovan Karadzic for nothing less than high crimes against humanity: for, in their callous disregard for the fatal consequences of their corrupt falsification of science, they are no less guilty of genocide than he.[55]

Viscount Monckton was one of a small number of deniers to attend the Copenhagen climate conference in December 2009. History will best remember his visit not because he proved at a stroke that the thousands of attendees in Copenhagen were wrong, but because Monckton called a group of young American protesters "Hitler Jugend." He also called NASA's James Hansen, "a fully-paid-up member of the new regime," saying he has "one of the unfailing hallmarks of Nazism and Fascism everywhere."[56]

Monckton is a ubiquitous figure in denial, seeming to pop up everywhere testifying, giving speeches, and debating. In the early weeks of 2010, Monckton descended on Australia. In one talk he allegedly accused NASA of destroying a satellite on launch rather than have the spacecraft fly, because if it were launched the satellite would prove that NASA was wrong and Monckton was right about global warming. Going on, he claimed he had discovered "a cure to a long-term illness that attacked his endocrine system." The drug had also "had positive results treating HIV and multiple sclerosis . . . colds [and] flu."[57]

Just as Howard Friel dissected Lomborg, so John Abraham, a well-published professor of mechanical engineering at the University of St. Thomas in Minnesota, has taken his microscope to a Monckton speech.[58] The indefatigable Abraham tracked each Monckton citation back to its source to determine whether his slides and remarks accurately represented the original work. In many cases Abraham wrote to the scientists who had produced the data that Monckton presented to see whether the scientists agreed with Monckton's description. They did not. As George Monbiot put it in a column in the *Guardian*, "Abraham's hard grind demonstrates that [Monckton's lecture] was a long concatenation of nonsense," with not one of his claims able to withstand scrutiny.[59]

Let us look at one Monckton claim and Abraham's rebuttal, a slide showing sea ice extent which Monckton titled, "Arctic Sea-ice Extent Is Just Fine: Steady for a Decade." Abraham wrote to scientists at three organizations which monitor Arctic sea ice; each disagreed with Monckton's account of their findings. One said, "Arctic sea ice has declined in all months, with the strongest downward trend at the end of the melt season in September."

Monckton, who has no academic credentials and who has never held an academic post or published in an academic journal, complained because Abraham did not "follow the usual practice in academe" of notifying Monckton in advance and simply asking for his references.[60] Abraham had questioned a slide sourced to "scienceandpublicpolicy. org," prompting Monckton to respond: "That, as he could have discovered if he had bothered—or, rather, dared—to check, was indeed the institution that had compiled the graph, taking the arithmetic mean of the global-temperature anomalies from the HadCrut, NCDC, RSS, and UAH datasets." But the Science and Public Policy Institute is not the source of original data, but a denier website. "This," Monckton concluded, "is the first of many indications of bad faith on Mr. Abraham's part that I shall be drawing to the attention of the authorities at the Bible College where he lectures."[61]

Monckton titled his speech at the 2009 Heartland Institute Conference, "Great Is Truth, and Mighty Above All Things." When asked by a reporter in 2007 whether he thought he had ever been wrong about anything at all, the Viscount Monckton of Brenchley replied, "Not on the big ones, no."[62]

American humorist Russell Baker once wrote, "Scratch any newspaperman and you'll find a man of letters who knows he could easily top . . . the best-seller charts if he chose to sacrifice the cocktail hour."[63] Baker, unwilling to make that considerable sacrifice, never did write his great novel. The relevance? One of the oddest things about the deniers is that they evidently believe that by "giving up the cocktail hour," they can discover facts unknown to an entire community of trained professionals armed with PhDs, supercomputers, and decades of experience. Will, Lomborg, and Monckton surely know, and Crichton surely knew, that for them to be right, the global community of scientists must be wrong. Do these deniers truly believe that they are smarter than, say, the entire collective membership of the American Geophysical Union, or that they have uncovered vital facts that have somehow escaped the notice of nearly every scientist in the world? Or do they believe that scientists as a class are so dishonest that they would promulgate global warming even though they know it to be false? For the deniers, there are only these two possibilities: either scientists are honestly wrong about global warming or they are lying, a charge I will address in chapter 13.

Toxic Tanks

The Think Tank has become so influential an institution in American politics that it seems almost to constitute another branch of government. Organizations like the American Enterprise Institute, the Brookings Institution, the Cato Institute, the Council on Foreign Relations, the Center for Economic and Policy Research, the Center for Strategic and International Studies, the Heritage Foundation, the Rand Corporation, and others less prominent, appear constantly in the media. According to Fairness and Accuracy in Reporting (FAIR), the top twenty-five U.S. think tanks, ranked by media citations, break down ideologically into 37 percent conservative, 47 percent centrist, and 16 percent progressive or left leaning.[1] Starting with the Reagan Revolution, the right-leaning think tanks helped to lay down the conservative theoretical foundation on such issues as supply-side economics and tax policy, immigration reform, affirmative action, welfare policy, and military defense. The Heritage Foundation largely wrote the Contract with America, most of whose provisions the Gingrich Congress adopted when it took office in 1994. These conservative think tanks provide a model, and sometimes actual support, for the global warming denier organizations that I profile in this chapter.

Of the many things the denier organizations have in common, the most obvious may be their admirable names. In its exposé of Exxon-Mobil's role in funding global warming denial (my subject in chapter 10), the Union of Concerned Scientists (UCS) published a list of forty-three organizations.[2] Not a single one has a name that reveals its opposition to global warming. Most sound eminently reasonable. At the top of the alphabetical list is Africa Fighting Malaria, which in 2004 received $30,000 for "climate change outreach." Fighting malaria in Africa—what a worthy cause! But on the organization's website one finds "an extensive collection of articles and commentary that argue against urgent action on climate change." Never mind that global warming may make malaria worse.

Or take the Frontiers of Freedom Institute. Who could oppose pushing freedom to and beyond its current frontiers? Between 1998 and 2005, the institute received over $1 million from ExxonMobil, over half of it designated for "climate change projects."

Harry and Louise

One of the first organizations to deny global warming was the Global Climate Coalition (GCC). Created in 1989 as it was becoming clear that the IPCC would be a major factor in the debate over global warming, the Global Climate Coalition operated from the offices of the National Association of Manufacturers. Supporting the Coalition were Amoco, the American Forest and Paper Association, the American Petroleum Institute, Chevron, Chrysler, Exxon, Ford, General Motors, Shell, Texaco, and the U.S. Chamber of Commerce. One of the coalition's first efforts was to hire the same public relations firm that pesticide companies had used to attack author Rachel Carson and her influential book, *Silent Spring*.

Prior to the Rio Earth Summit, the Global Climate Coalition distributed a video claiming that far from being a problem, more atmospheric carbon dioxide would merely increase crop production and help feed the world. The Coalition fought the Kyoto Protocol through an advertising campaign that employed the same public relations firm that produced the "Harry and Louise" ads that helped defeat the Clinton health plan proposal in the 1990s. Of the Kyoto Protocol, the ads claimed "It's not global and it won't work," alleging that curtailing carbon emissions would cost Americans $.50 more per gallon of gasoline.[3]

By 1997 the IPCC had released its second report. Some business leaders had evidently begun to realize that continuing to deny global warming, which their own scientists were telling them was real, would ultimately be bad for business. John Browne, chairman of British Petroleum, said in a speech at Stanford University that, "The time to consider the policy dimensions of climate change is not when the link between greenhouse gases and climate change is conclusively proven, but when the possibility cannot be discounted and is taken seriously by the Society of which we are a part. We and BP have reached that point." In 1999 the young chairman of Ford Motor Company, William Clay Ford Jr., great grandson of Henry Ford, said, "I expect to preside over the demise of the internal combustion engine." Explaining why he pulled Ford out of the Global Climate Coalition, Ford said, "The present risk is clear. The climate appears to be changing, the changes appear to be outside natural variation, and the likely consequences will be serious. From a business planning point of view, that issue is settled. Anyone who disagrees is, in my view, still in denial."[4]

In 1997, Amoco, British Petroleum, DuPont, Ford, and Shell resigned from the Global Climate Coalition to join a progressive new group, the Business Environmental Leadership Council, founded by the Pew Center on Global Climate Change. The two core principles of the group are

- We accept the views of most scientists that enough is known about the science and environmental impacts of climate change for us to take actions to address its consequences, and
- Businesses can and should take concrete steps now in the United States and abroad to assess opportunities for emission reductions, establish and meet emission reduction objectives, and invest in new, more efficient products, practices and technologies.[5]

Big Oil and Big Auto were taking an obvious lesson from Big Tobacco, whose denial of a link between smoking and health had caused a loss of public confidence and respect that began to influence court decisions and the success of plaintiffs' claims. The exposure of Big Tobacco's lies figured in its agreement to pay state governments $251 billion to compensate them for the Medicare costs of treating smoking-induced health problems. The trial led to the release of the documents that make up the Legacy Tobacco database. If one wants a definition of "Big," call it the ability of an industry to shell out over $250 billion without going broke.

The Global Climate Coalition responded to the corporate withdrawals with a "strategic restructuring" so that its members would comprise trade associations, rather than individual companies, and thus have deniability. But it didn't work, for after the IPCC's Third Assessment report appeared in 2001, the GCC disbanded because it had "served its purpose by contributing to a new national approach to global warming." The statement continued, "The Bush administration will soon announce a climate policy that is expected to rely on the development of new technologies to reduce greenhouse emissions, a concept strongly supported by the GCC."[6] Thus after spending more than a decade denying that increased greenhouse gas levels were even a problem, in a deathbed conversion the Global Climate Coalition reversed itself and endorsed the false hope put forth by the Bush administration that yet-to-be-invented technologies will solve the problem of global warming.

Tale from the Crypt

In 2007 a court action in California resurrected an important and long-buried set of energy-company documents. Big Auto had sued the state over its intent to restrict automobile emissions beyond the level called for in federal standards. In the legal discovery process, a report surfaced that Mobil Oil Company chemical engineer L. C. Bernstein had written in 1995 on behalf of the Science and Technology Assessment Committee, a group of industry scientists formed to advise the Global Climate Coalition.[7] The report went to thirty-five oil and coal companies, to electric utilities, attorneys, the National Mining Association, and so on, enough organizations to ensure that the committee's findings must have been common knowledge among the energy companies and others that were then denying global warming. Yet just as Big Tobacco spent millions trying to convince the public that smoking was not harmful to health, even though it knew just the opposite to be true, so the companies supporting the Global Climate Coalition continued to pour millions into convincing the public that global warming was not dangerous, even though their own scientists were telling them it was.

The Bernstein report began by noting that the latest IPCC Summary for Policymakers had concluded that humans were influencing the climate. Bernstein and his committee would not go that far, saying it "believes that the IPCC statement [of 1995] goes beyond what can be justified by current scientific knowledge." Then came the shocker: "The scientific basis for the greenhouse effect and the potential impact of human emissions of greenhouse gases such as CO_2 on climate is well established and cannot be denied." The report dispensed with the argument that "It's the Sun": "Over the past 120 years, the maximum contribution of solar variations to global temperatures would be 0.1%, about one-fifth of the temperature change actually observed during that period." During the 1990s, satellite temperature measurements had seemed to call global warming into question, but the industry scientists found otherwise: "The corrected satellite measurements still do not agree with a land-based temperature record, but they both show warming."

The report went on to debunk the contrarian views of Patrick Michaels and others, saying, "These alternative hypotheses do not address what would happen if atmospheric concentrations of greenhouse gases continue to rise at projected rates." In other words, whether or not human activities have caused the observed rise of carbon dioxide since the Industrial Revolution, elementary physics demands that continued burning of fossil fuels will increase the levels of carbon dioxide in the atmosphere and cause more warming.

The report wound up: "The contrarian theories raise interesting questions about our total understanding of climate processes, but they do not offer convincing arguments against the conventional model of greenhouse gas emission-induced climate change."

Please Don't Poop in My Salad

On the Heartland Institute website, a banner flashes portraits of famous thinkers: Benjamin Franklin, Thomas Jefferson, John Locke, James Madison, Thomas Paine, Joseph L. Bast . . . Joseph L. Bast?[8] Who is he and what the heck is he doing in this distinguished lineup? It turns out Bast helped found the Heartland Institute in 1984 and as of 2011 still serves as its president. His best-known work is *Please Don't Poop in My Salad*, a collection of essays about freedom, especially the freedom to smoke whenever, wherever, and as much as you like at the lowest possible tax per pack.

The institute's mission is "To discover, develop, and promote free-market solutions to social and economic problems." These include "parental choice in education, choice and personal responsibility in health care, market-based approaches to environmental protection, privatization of public services, and deregulation in areas where property rights and markets do a better job than government bureaucracies." More specifically, the institute favors "common-sense environmentalism." The Heartland Institute opposed the Kyoto Protocol and favors genetically engineered crops, school vouchers, and deregulating the health care insurance industry.

The institute says it is "a genuinely independent source of research and commentary," yet since its founding the Heartland Institute has been supported by and has itself supported Big Tobacco. For years, Heartland had a symbiotic relationship with Philip Morris, source of a 1995 memo saying that the company used its philanthropic contributions "as a strategic tool to promote our overall business objectives and to advance our government affairs agenda." The company did so in particular by supporting "the work of free market 'think tanks' and other public policy groups whose philosophy is consistent with our point of view." Among the think tanks listed was the Heartland Institute.[9]

Until May 2003, the manager of industry affairs for Philip Morris, Roy E. Marden, was a Heartland board member. The institute sponsored a conference on behalf of Philip Morris "on the impact of federal mandates/EPA regulations" as part of the company's response to the EPA decision to classify secondhand smoke as a proven lung carcinogen.

In 2006 the Heartland Institute partnered with the National Association of Tobacco Outlets in a campaign to change public opinion in order to "prevent . . . statewide smoking bans." Bast jointly wrote an essay titled "Tobacco and Freedom," arguing that smokers already pay too much tax, that attempts to restrict tobacco and smoking are based on junk science, that consumers who sue tobacco companies are engaged in "lawsuit abuse," and that "punishing smokers 'for their own good' is repulsive to the basic libertarian principles that ought to limit the use of government force."[10] Says Bast, "Several experts have told me that smoking fewer than seven cigarettes a day does not raise the smoker's risk of lung cancer."[11]

SourceWatch reports that between 1998 and 2006, the Heartland Institute received $676,500 from ExxonMobil. We do not know how much it received after 2006, for the institute stopped identifying its donors. Bast explained: "For many years, we provided a complete list of Heartland's corporate and foundation donors and challenged other think tanks and advocacy groups to do the same. To our knowledge, not a single group followed our lead. Critics . . . found the donor list a convenient place to find the names which they used in *ad hominem* attacks against us."[12]

A recent institute publication is *The Skeptic's Handbook*, a slim volume that Heartland is distributing to more than 150,000 people across the country, including 850 journalists, 26,000 schools, and 19,000 leaders and politicians. Funded by an "anonymous donor," the largest single recipient of the handbook will be Black churches, who will receive 25,962 copies. Trustees at colleges and universities will run a close second, receiving 20,253 copies.[13] The book is available for downloading at the institute's website.

The handbook's author is Joanne Nova, a stage surname, whose most recent activity was to tour Australia with a science program sponsored by Shell Oil. A "professional speaker," Nova's one previous publication was *Serious Science Party Tricks*. At the March 2009 Heartland Institute Conference, she gave a talk titled, "The Great Global Fawning: How Science Journalists Pay Homage to Non-Science and Un-Reason."

Nova's website says that, "Donors have paid for over 160,000 copies so far in the U.S., Australia, New Zealand, Sweden, and soon in Germany. Over 60,000 copies have been downloaded from this site (and countless others from copies on other sites). Plus volunteers have translated it into German, French, Norwegian, Finnish, Swedish, Turkish, Portuguese, Danish, Japanese, Balkan, Spanish, Thai, Czech and Lao. The second *Skeptic's Handbook* is available in French and Turkish. (Versions in Dutch, and possibly Italian are on the way)."

The cover of the handbook says, "The only thing that matters here is whether adding more CO_2 to the atmosphere will make the world much warmer," a question that Arrhenius answered in 1896. The second page goes on, "Only four points [are] worth discussing. Every time you allow the conversation to stray, you get stuck in a dead-end and miss the chance to definitively expose the lack of evidence that carbon is 'bad.'"

Each of the four points has been claimed by deniers, debunked by scientists, claimed again by deniers, debunked again, and now shows up once more in *The Skeptic's Handbook*. In the following numbered section, the first paragraph, enclosed in quotation marks, comes from the handbook (bold in the original); the second is my response.

1. "**The greenhouse signature is missing**. Weather balloons have scanned the skies for years but can find *no sign* of the telltale 'hot spot'

warming pattern that greenhouse gases would leave. There's not even a hint. *Something else caused the warming.* If we can't get good results from a simple weather balloon, what chance do we have with a computer model? [italics in original]."

Climate models predict that the troposphere, the layer of the atmosphere closest to the earth, should warm faster than the surface. For a number of years, weather balloons have been unable to measure the warming and confirm the prediction. Nova calls this the "knockout blow." How could scientists have missed it? They didn't. In a paper in the peer-reviewed *International Journal of Climatology* in November 2008, a consortium of scientists from twelve different institutions reported that there was no fundamental discrepancy between modeled and observed tropical temperature.[14] They say that claims to the contrary were due to a flawed statistical test and the use of older observational datasets.

2. **"The strongest evidence was the ice cores, but newer, more detailed, data turned the theory inside out.** Instead of carbon pushing up temperatures, for the last half-a-million years temperatures have gone up *before* carbon dioxide levels. On average 800 years *before*. This totally threw what we thought was cause-and-effect out the window. *Something else caused the warming.*"

Scientists know that small, repeating changes in the earth's orbit and axial tilt—the Milankovitch cycles—initiate warming, after which the greenhouse effect and its feedbacks kick in to amplify warming. Thus the fact that temperatures rise first, followed immediately by CO_2 feedbacks, is perfectly well understood.

3. **"Temperatures are not rising.** Satellites circling the planet twice a day show that the world has not warmed since 2001. How many more years of NO global warming will it take? While temperatures have been flat, CO_2 has been rising, BUT *something else has changed the trend.* The computer models don't know what it is."

As discussed several times, the year 1998 saw both a major El Niño and record temperatures. By 2008 a La Niña, which tends to lower temperatures, had settled in. Plus sunspot activity was minimal during the period. By the summer of 2009, NASA forecast the arrival of a new El Niño and summer sea surface temperatures were the warmest on record. The year 2009 was the warmest on record in the Southern Hemisphere and tied for second warmest globally. The year 2010 tied 2005 as the warmest on record. Temperatures *are* rising.

4. **"Carbon dioxide is already doing almost all the warming it can do.** Adding twice the CO_2 doesn't make twice the difference. The first CO_2 molecules matter a lot, but extra ones have less and less effect. In fact, carbon levels were ten times as high in the past but the world still slipped into an ice age. Carbon today is a bit-part player."

As one blogger pointed out, this is like "suggesting that throwing more wood on a fire will not make it bigger."[15] It is true that the relationship between temperature and carbon dioxide concentrations is not a straight line. That is why scientists focus on climate sensitivity—the temperature increase that doubling carbon dioxide concentrations will cause, currently estimated at $3°C$ ($5.4°F$). If "carbon dioxide is already doing almost all the warming it can," why have carbon dioxide concentration and temperatures risen during the second half of the last century, and why are both continuing to rise? Scientists know why; Nova and the Heartland Institute do not know.

Page 7 of the handbook reveals that "the main 'cause' of global warming is air conditioners," which give off heat and are located too close to temperature measuring stations. The National Oceanographic and Atmospheric Administration thoroughly debunked that claim, which in any case does not account for higher temperatures in parts of the world without air conditioners, or in temperatures measured by satellites.

Embarrassing General Marshall

The oldest anti–global warming organization is the George C. Marshall Institute, founded in 1984 by members of the Jasons, the behind-the-scenes group of scientists who advised the Pentagon about various matters, including whether to use tactical nuclear weapons in Vietnam. At this writing, the current president of the Marshall Institute is William O'Keefe, formerly employed by the American Petroleum Institute. The 2007 exposé by the Union of Concerned Scientists identified Stephen Milloy and Frederick Seitz as Marshall Institute "key personnel" who participated both in the Big Tobacco and the climate denial "disinformation campaigns."[16] The UCS lists eleven "scientific spokespersons" who are affiliated with ExxonMobil-funded front groups, seven of them with the Marshall Institute: Sallie Baliunas, Sherwood Idso, David

Legates, Richard Lindzen, Patrick Michaels, Frederick Seitz, and Willie Soon.

Between 1998 and 2005, the Marshall Institute received $630,000 from ExxonMobil, as well as additional funding from conservative private foundations such as the Sarah Mellon Scaife and John M. Olin Foundations. Like the Heartland Institute, the Marshall Institute no longer publishes its donor list.

In the 1980s, led by some of the Jasons, the Marshall Institute took a prominent role in supporting President Ronald Reagan's Strategic Defense Initiative, aka "Star Wars," which many scientists opposed. Once SDI faded from public debate, the institute took up global warming, releasing in 1989 a report claiming that "cyclical variations in the intensity of the sun would offset any climate change associated with elevated greenhouse gases."[17] The IPCC and the Bernstein report, described in the last chapter, both debunked the claim that "It's the Sun," but it refuses to die.

Two of the scientists associated with the Marshall Institute, Baliunas and Soon, work at the Harvard-Smithsonian Center for Astrophysics. In 2003 the two published a review article in a journal called *Climate Research*, alleging that there was nothing special about twentieth-century temperatures: "Across the world, many records reveal that the 20th century is probably not the warmest nor a uniquely extreme climatic period of the last millennium."[18] By the time the article appeared, the original "hockey stick" paper (discussed in chapter 12) had been out for four years and was doing a lot of damage to the denier cause. The Marshall Institute made the Baliunas-Soon paper available on its website; Senator Inhofe trumpeted it as evidence that the "overwhelming factor" in climate change was not human activity but the Sun.

Three of the editors of *Climate Research*, including the incoming editor in chief, resigned in protest over the journal's publication of the Baliunas-Soon article. The editor in chief was concerned that "Some of the skeptics had identified *Climate Research* as a journal where some editors were not as rigorous in the review process as is otherwise common," adding that he regarded the manuscript as "flawed." Moreover, thirteen of the scientists whom Baliunas and Soon had cited in the article published a rebuttal claiming that the pair had seriously misinterpreted their research. Jeff Nesmith of Cox News Service discovered that the American Petroleum Institute had funded the Baliunas-Soon study.[19]

The resignations and rebuttal had no visible effect on the careers of Baliunas and Soon, who remain among the deniers' favorite scientists. Baliunas has been associated with nine different denier organizations that ExxonMobil has funded, including the Marshall Institute, which lists her as one of its "experts" and which in 1997 paid her $52,000 for serving as a director.[20] That same year, Baliunas won the Petr Beckmann Award for her "devastating critique of the global warming hoax."[21] Doctors for Disaster Preparedness, another admirably named organization, which has close ties to the Oregon Institute of Science and Medicine, gives the award for "courage and achievement in defense of scientific truth and freedom." Winners of the Beckmann award are a Who's Who of denial, including Sherwood Idso, Robert Jastrow, Arthur B. Robinson, S. Fred Singer, and Baliunas's coauthor, Willie Soon.[22]

Let senator and presidential candidate John McCain have the last word on the George C. Marshall Institute. In 2004 the Arctic Climate Impact Assessment included the chart (fig. 6.1) shown on page 52, showing how temperatures, carbon dioxide concentrations, and fossil fuel emissions have all increased together. The Marshall Institute went on the attack, claiming that the report used "unvalidated climate models and scenarios . . . that bear little resemblance to reality and how the future is likely to evolve." McCain responded, "General Marshall was a great American. I think he might be very embarrassed to know that his name was being used in this disgraceful fashion."[23]

Push Down on the Accelerator

"One of Washington's feistiest think tanks," writes the *Boston Globe*. "The best environmental think tank in the country," says the *Wall Street Journal*.[24] Founded by Fred L. Smith Jr. in 1984, Competitive Enterprise Institute (CEI) has over forty policy experts and staff who are quick off the mark on any issue involving government regulations. As the institute's website says, "If the government has rewritten the rules to regulate it, we have an opinion how it could be done better."

True to its motto, CEI has opposed government regulation in just about every field one can imagine: air quality standards, dioxin, drug safety, fuel efficiency standards, labeling of alcoholic beverages, rent control, securities law, and more. CEI protested the publication of a book called *Our Stolen Future* that identified ways in which chemical pollutants cause birth defects, sexual abnormalities, and reproductive failures. In the late 1990s, CEI filed lawsuits challenging the constitutionality of the 1998 Big Tobacco Master Settlement Agreement. Keeping up with the times, the organization is now opposing regulation of high technology, e-commerce, intellectual property, telecommunications, and the like.

CEI has received funding from Amoco, Philip Morris, and especially from ExxonMobil, which between 1998 and 2005 gave it over $2 million. In 1992, CEI's budget came to $765,000. The money helped fund a PR program called "Earth Summit Alternatives," which arranged media interviews and generated articles opposing the results of the Rio Summit. In 1997, CEI contested the outcome of the Kyoto climate negotiations, offering to provide experts who would promote the claim that "global warming is a theory not a fact," the same strategy that creationists use to deny evolution. Also in 1997, CEI helped create the Cooler Heads Coalition, "to dispel the myths of global warming by exposing flawed economic, scientific and risk analysis." It offered CEI speakers who were prepared to explain five major points:

- Many scientists are skeptical of climate change theory.
- A warm Earth is a prosperous Earth.
- The Kyoto Protocol will not substantially reduce greenhouse gas emissions worldwide.
- Global warming policies will harm the U.S. economy.
- Preventing global warming could undermine U.S. national security and global economic health.

Among the CEI "experts" ready to make these points were Sallie Baliunas, Patrick J. Michaels, and S. Fred Singer.

In 1998 a CEI official argued before a committee of the U.S. House of Representatives that, "Where per capita energy consumption is high, per capita income is also high; and where per capita energy consumption is low, per capita income is also low. Thus if we are to rescue mankind from the perils of poverty, we must dramatically increase global energy consumption. We must push down on the accelerator."[25] (This claim may be generally true, though the oil emirates with their high energy consumption but relatively low GDP show there are exceptions. But it does not follow that high energy consumption <u>causes</u> high GDP.)

Between May 18 and May 28, 2006, CEI ran television ads in four-teen U.S. cities in an effort to counter the publicity generated by Al Gore's film, *An Inconvenient Truth*. One said, "The Antarctic ice sheet is getting thicker, not thinner. . . . Why are they trying to scare us?" One of the mainstream scientists whose research the ad quoted called it a "deliberate effort to confuse and mislead the public about the global warming debate," accusing CEI of "selectively using only parts of my previous research to support their claims. They are not telling the entire story to the public."[26] Of course, his protest reached only a tiny fraction of those who saw the ads. The second ad used another deceptive denier argument, showing a little girl blowing a dandelion while a narrator said of carbon dioxide, "They call it pollution; we call it life."

One case study involving CEI shows that with a sympathetic president occupying the White House and his party in control of Congress, de-niers can reach all the way into the Oval Office and even secure the firing of a Cabinet officer. The Global Change Research Act, which Congress passed in 1990, mandated that the United States produce a National Assessment of Climate Change. The first assessment began under the Clinton administration and included twenty study teams based at uni-versities. Only about half the work of the assessment focused on science; the other half aimed at educating local stakeholders, including farmers, fishermen, businessmen, mayors, and the like. The pathbreaking effort led to a 150-page report and an accompanying 600-page background document. To provide more than a single projection of what global warming might bring, the first assessment used two different global climate models, both of them relatively primitive by today's standards.

When the report appeared in 2000, near the end of the Clinton ad-ministration, Republicans mounted a fierce effort to discredit it. CEI, with Senator Inhofe as a coplaintiff and William Jefferson Clinton as defendant, filed a lawsuit in the U.S. District Court alleging that the report's preparers had not followed proper procedures. The suit de-manded that the government refrain from printing or using the report. One of the coplaintiffs, Rep. Jo Ann Emerson (R-MO), said that, "The administration is rushing to release a junk science report in violation of current law to try to lend support to its flawed Kyoto protocol negotia-tions." The assessment, Emerson said, was "a biased, doom and gloom piece of science fiction."[27]

George W. Bush had campaigned as an environmentalist, saying that once in office he would act to regulate carbon dioxide emissions, a pledge that did not survive the first few months of his administration. Nevertheless, an international treaty obligated Bush to file with the United Nations a "United States Climate Action Report," in whose preparation it was natural to draw on the National Assessment. Instead of fighting the CEI lawsuit, the Bush administration surrendered, saying that the assessment's conclusions were "not policy positions or official statements of the U.S. government. Rather, they were produced by the scientific community and offered to the government for its consideration."[28]

ExxonMobil objected to the report's finding that climate change is dangerous and caused by humans. Big Oil and CEI had an ally in the White House, a man named Philip Cooney who had worked for the American Petroleum Institute. At the time the assessment appeared, Cooney was chief of staff of the White House Council on Environmental Quality.

Cooney wrote to Myron Ebell of the Competitive Enterprise Institute, asking for the organization's help in deep-sixing the National Assessment report. Ebell advised Cooney to keep President Bush as far as from the report as possible. Within days, Bush dismissed the assessment as "put out by the bureaucracy."[29]

In September 2003, Ebell wrote to Cooney:

We made the decision this morning to do as much as we could to deflect criticism by blaming the EPA for freelancing. It seems to me that the folks at the EPA are the obvious fall guys, and we would only hope that the fall guy (or gal) should be as high up as possible. Perhaps tomorrow we will call for [EPA head Christine] Whitman to be fired.[30]

One year later, Whitman had departed the Bush cabinet.

CEI's point of attack was the assessment's use of two different climate models. Patrick Michaels testified before Congress on July 25, 2002: "Under the ethics of science, [models] should have been abandoned or modified, rather than used as input to a document with substantial policy implications." The National Assessment, Michaels said, "should be rejected from the public record."[31]

In August 2003, CEI brought a second lawsuit against the National Assessment, this time using a new vehicle, the Data Quality Act, which Representative Emerson had helped to write and which was a key front of the Republican war on science, to co-opt the title of the essential book by Chris Mooney.[32] Listed as defendant was George Walker Bush. Once again the White House capitulated, saying on the government's National Assessment website that the report had not been subject to Data Quality Act guidelines, giving the impression that it was unreliable. Yet at the time the report appeared, the Data Quality Act did not yet exist.

Rick Piltz was a senior associate in the United States Climate Change Science Program, which had oversight of the National Assessment. In March 2005, Piltz resigned over White House interference with the assessment report. He went on to create Climate Science Watch, part of the Government Accountability Project, an agency to protect whistleblowers. Piltz reported that

> The Bush administration has essentially buried [the assessment] by refusing to discuss it and has directed federal climate science program leaders in the agencies to refrain from any substantive reference to or use of the National Assessment in public statements, reports to Congress, and research planning. The administration disbanded the developing networks of scientists and stakeholders and has refused to initiate a follow-on second National Assessment of climate change impacts.[33]

Piltz went on to show how the Bush administration's Climate Change Strategic Plan had referred to the National Assessment twelve times in its first draft, but in the final draft only in a single sentence that did not even include the report's title.

The National Academy of Sciences praised the assessment and criticized the administration for burying it. In 2005, ten scientists, half of them not previously associated with the assessment, conducted a survey and held a workshop to evaluate the assessment report. They concluded: "the assessment was largely successful in implementing its basic design of distributed stakeholder involvement and in achieving its basic objectives."[34] Senators Kerry and McCain asked the Government Accounting Office to investigate whether the strategic plan met the requirements of the Global Change Research Act. GAO reported that the plan had properly considered current environmental trends and projected future outcomes, "but fell short when it came to analyzing the effects of climate change on natural resources, biodiversity, agriculture, and other impact areas directly relevant to humans."[35] In other words, the GAO found that the report did not go far enough in examining the consequences of global warming.

In 2006 the Center for Biological Diversity, Greenpeace, and Friends of the Earth together sued the Bush administration for its failure to issue the National Assessment as the law required. In August 2007 a judge agreed and ordered a new climate plan by March 1, 2008, and a new assessment by May 31, 2008. The Bush administration met the schedule, but the result was mainly a literature review.

The Climate Change Science Program under the administration of President Barack Obama has made available a DVD of the original, squelched 2000 National Assessment, which though dated contains much useful information. The program also created a new website and released a new report: *Global Climate Change: Impacts in the United States*.[36]

The Competitive Enterprise Institute's point person on global warming is Myron Ebell, a philosophy major at the University of California at San Diego who studied political theory at the London School of Economics and history at Cambridge University. In a 2007 interview in *Vanity Fair*, Ebell said that the "hockey stick" is wrong; CO_2 levels used to be far higher, so there is no cause for alarm today; the earth has entered a cooling phase comparable to the Little Ice Age and the CO_2 we are emitting is merely maintaining the earth as a pleasant place to live; the oceans are not warming; computer-based climate models "don't even pass the laugh test"; most of the big storms that kill people occur in winter; warming is not causing animal habitats to shift; global warming does not threaten polar bears; the data showing Arctic warming are "cooked"; the bark beetle infestations that are killing millions of trees in western North America are merely due to "poor forest management, not global warming."[37]

The best way to deal with global warming, Ebell recommends, is to burn more fossil fuel. You see, the more prosperous the energy industry, the more quickly it can replace its fossil fuel plants with new technology that does a better job of trapping greenhouse gases. To the extent that global warming is a threat, we should wait until we can tackle it with more intelligence and efficiency and at less cost.

Like other deniers, philosophy major Ebell reserves special venom for James Hansen, who he says "was not trained as a climate scientist [but] as an astronomer. He's a physicist," he continues, whose "dissertation was on the atmosphere of Venus, and he has applied what he's learned in physics and in astronomy to become a climate scientist." Of himself, Ebell says, "I'm not claiming to be a climate authority—the way Jim Hansen is. I'm just giving you the informed layman's perspective."

An Industry to Trust

The industry of denial has spent millions of dollars in its campaigns. Most of its front organizations are public charities that have little or no funds of their own. Where did they get the money? Thanks to IRS reporting requirements, we know the answer and one source dwarfs all others: ExxonMobil Corporation.[1] Between 1998 and 2005, the company gave $16 million to more than forty organizations that deny global warming. The American Enterprise Institute (AEI) received $1,625,000; Lee R. Raymond, ExxonMobil's chairman and CEO, served as vice chair of AEI's Board of Trustees. In February 2007, AEI wrote to scientists and economists in Britain and the United States offering each $10,000 for articles criticizing the IPCC's Fourth Assessment, which was about to appear.[2] The letter accused the IPCC of being "resistant to reasonable criticism and dissent and prone to summary conclusions that are poorly supported by the analytical work." The essays were to "thoughtfully explore the limitations of climate model outputs."

Most Profitable Company in History

Over the 1998–2005 period, ExxonMobil gave the Annapolis Center for Science-Based Public Policy, where Harrison Schmitt served as president, $763,500, 20 percent of the center's budget. A little-known organization, Citizens for a Sound Economy Educational Foundation, received $380,250, allowing it to claim that the science of climate change is "far from settled," citing as its authority Sallie Baliunas.[3] The Fraser

Institute, home to "hockey stick" denier Ross McKitrick, received $120,000; the George C. Marshall Institute, $630,000; the Heartland Institute, $676,500; the Heritage Foundation, $460,000; the National Center for Policy Analysis, which published a paper claiming that global warming did not threaten the Arctic polar bear, $420,900; Tech Central Station, a denier website run by an ExxonMobil lobbying firm, $95,000. And so on.

ExxonMobil is the world's largest publicly traded company. Its 2005 revenues exceeded the gross domestic product of all but a handful of nations. According to one report, ExxonMobil is the most profitable company in the history of the world, generating almost $100 million in profits—each day.[4] The amount of carbon dioxide emitted through the use of ExxonMobil products would rank the company sixth among polluting nations, just after India, before Germany, and well ahead of Canada and the United Kingdom. ExxonMobil has a lot at stake.

In its exposé titled *Smoke, Mirrors, and Hot Air: How ExxonMobil Uses Big Tobacco's Tactics to Manufacture Uncertainty on Climate Science*, the Union of Concerned Scientists accused ExxonMobil of using its profits to mount "the most sophisticated and successful disinformation campaign since big tobacco misled the public about the incontrovertible scientific evidence linking smoking to lung cancer and heart disease."[5] The report pointed out that not only had ExxonMobil used the same tactics as the tobacco companies, it used some of the same people: Stephen Milloy and Frederick Seitz, to name two. Like the tobacco companies, ExxonMobil manufactured uncertainty, laundered information, promoted pseudoscience, called for "sound science" to divert attention from incontrovertible scientific evidence, and used its access to government to deny and delay action.

By 1997 the Global Climate Coalition, described in the previous chapter, had begun to crumble as British Petroleum, Shell, and Texaco pulled out. ExxonMobil took the opposite tack, ramping up its campaign of denial. The first step was to invent and underwrite a new group, the Global Climate Science Team, which included Jeffrey Salmon of the Marshall Institute; Myron Ebell, then with Frontiers of Freedom Institute; Randy Randol of ExxonMobil; Stephen Milloy of the Advancement of Sound Science Coalition (funded by Philip Morris); and Joseph Walker of the American Petroleum Institute, who served as the coordinator of the committee. In April 1998, Walker presented the team with an action plan that defined success in a section titled "Victory Will Be Achieved When"

- "Average citizens 'understand' [recognize] uncertainties in climate science; recognition of uncertainties becomes part of the 'conventional wisdom.'"
- "Media 'understands' [recognizes] uncertainties in climate science."
- "Media coverage reflects balance on climate science and recognition of the validity of viewpoints that challenge the current 'conventional wisdom.'"
- "Industry senior leadership understands uncertainties in climate science, making them stronger ambassadors to those who shape climate policy."
- "Those promoting the Kyoto treaty on the basis of extent [sic] science appears [sic] to be out of touch with reality."[6]

Walker recommended that ExxonMobil and its public relations firm "develop and implement a program to inject credible science and scientific accountability into the global climate debate, thereby undercutting the 'prevailing scientific wisdom.'"[7]

ExxonMobil began to support not only large think tanks like the American Enterprise Institute and the Cato Institute but also a host of smaller and lesser-known organizations, some apparently created just to receive ExxonMobil support. But why did ExxonMobil spread its money so thinly? Why give nearly $16 million to more than forty organizations, rather than choosing a smaller set of the most effective and giving each a larger share? The answer may be that the more admirably named organizations that question the consensus on global warming, the greater the doubt that can be created in the public's mind.

Soon after the 2000 presidential election, ExxonMobil was at the table with the secret Energy Task Force that Vice President Dick Cheney assembled to advise the administration about energy policy. The U.S. chief negotiator in climate meetings in Buenos Aires and Montréal told ExxonMobil's lobbyist and other members of the Global Climate Coalition that President Bush had "rejected Kyoto, in part, based on input from you."[8]

We know from discoveries under the Freedom of Information Act that after the 2001 IPCC report appeared, an ExxonMobil lobbyist succeeded in ousting the U.S. chair of the IPCC, Dr. Robert Watson.[9] The lobbyist, Randy Randol, complained in a letter to the White House that Watson had been "hand-picked by Al Gore," and urged Watson's replacement. Watson's sin? He believed and said that human-induced global warming was real. Here was heresy, and when Watson's term as IPCC chair ended, the Bush administration declined to renominate him for a second term. Rajendra Pachauri replaced Watson, who now serves as Chief Scientist of the World Bank. But Pachauri turned out to be a staunch defender of global warming science, leading the deniers to accuse him of having a conflict of interest in receiving funding from oil and power companies, a charge that Pachauri denies.

As noted earlier, to serve as Chief of Staff in the White House Council on Environmental Quality, George W. Bush hired a former employee of the American Petroleum Institute named Philip Cooney. At API, Cooney had been "climate team leader," charged with preventing the United States from adopting legislation or entering into an international agreement that might restrict carbon emissions. According to whistleblower Rick Piltz, Cooney radically altered the language of a Climate Change Science Program document. For example, the official draft had said that

Warming will also cause reductions in mountain glaciers and advance the timing of the melting of the mountain snow packs in polar regions. In turn, runoff rates will change and flood potential will be altered in ways that are currently not well understood. There will be significant shifts in the seasonality of runoff that will have serious impacts on native populations that rely on fishing and hunting for their livelihood. These changes will be further complicated by shifts in precipitation regimes and a possible intensification and increased frequency of extreme hydrologic events.

Cooney struck that entire section on the grounds that it was "straying from research strategy into speculative findings/musings." Two days after the *New York Times* first reported Piltz's whistleblowing, Cooney resigned his White House office. One week later, he was a high-ranking public relations officer at ExxonMobil.[10]

According to SourceWatch, Congressman Joe Barton, chair of the House Energy and Commerce Subcommittee from 2004 through 2006, has received well over $1 million from the oil and gas industry over his career. Sen. James Inhofe received $662,506 from oil companies between 2000 and 2008, making him the number one recipient of oil money.

Open opposition to ExxonMobil's role in the campaign of denial broke out in 2006. In October of that year, Senators Olympia Snowe and Jay Rockefeller wrote to ExxonMobil chairman and CEO Rex Tillerson saying:

In light of the adverse impacts still resulting from your corporation's activities, we must request that ExxonMobil end any further financial assistance or other support to groups or individuals whose public advocacy has contributed to the small, but unfortunately effective, climate change denial myth.[11]

That same autumn, Bob Ward, then senior manager of policy communication of the Royal Society, Britain's equivalent of the U.S. National Academy of Sciences, wrote to Nick Thomas, director of corporate affairs for ExxonMobil's UK division.[12] Ward had been surprised to find in the 2005 ExxonMobil Corporate Citizenship Report the following paragraph:

While assessments such as those of the IPCC have expressed growing confidence that recent warming can be attributed to increases in greenhouse gases, these conclusions rely on expert judgment rather than objective, reproducible statistical methods. Taken together, the gaps in the scientific basis for theoretical climate models and the interplay of significant natural variability make it very difficult to determine objectively the extent to which recent climate changes might be the result of human actions.[13]

Ward reminded Thomas that the expert judgment of the IPCC "was actually based on objective and quantitative analyses of methods, including advanced statistical appraisals, which carefully accounted for the interplay of natural variability, and which had been independently reproduced." Ward found it "very difficult to reconcile the misrepresentations of climate change science in these documents with ExxonMobil's claim to be an industry leader."

Ward had reviewed ExxonMobil's 2005 Worldwide Giving Report and counted thirty-nine organizations that promoted misinformation about the science of global warming, "by outright denial of the evidence that greenhouse gases are driving climate change, or by overstating the amount and significance of uncertainty in knowledge, or by conveying a misleading impression of the potential impacts of anthropogenic climate change." ExxonMobil had provided these organizations with more than $2.9 million. Ward reminded Thomas that ExxonMobil had pledged not to give any further support to these organizations and asked the company to inform the climate scientists who are Fellows of the Royal Society whether it "will be continuing to express views that are inconsistent with the findings of their work."

In May 2008, ExxonMobil pledged in its Corporate Citizenship Report that, "We will discontinue contributions to several public policy research groups whose positions on climate change could divert attention from the important discussion on how the world will secure the energy required for economic growth in an environmentally responsible manner."[14]

Ward, now policy and communications director at the Grantham Research Institute on Climate Change and the Environment at the London School of Economics, pointed out in an article in the *Guardian*: "ExxonMobil has been briefing journalists for three years that they were going to stop funding these groups. The reality is that they are still doing it. If the world's largest oil company wants to fund climate change denial then it should be upfront about it, and not tell people it has stopped."[15]

As Ward surely knows, when you read anything written by Big Oil, parse it carefully. First of all, the above quotation from the Citizenship Report does not say anything about the reality and danger of global warming, other than that it could "divert attention from "secur[ing] the energy for economic growth." More importantly, "several" does not mean "all": ExxonMobil's list of grantees for 2008 includes $76,106 to the Smithsonian Astrophysics Observatory, home to Sallie Baliunas and Willie Soon, authors of the discredited *Climate Research* article, and $105,000 to the Annapolis Center for Science-Based Public Policy, which lists Baliunas on its Advisory Council. The 2009 giving report shows the two organizations receiving the same amount a year later.

ExxonMobil's website says:

There is increasing evidence that the earth's climate has warmed on average about 0.7°C in the last century. Many global ecosystems, especially the polar areas, are showing signs of warming. CO_2 emissions have increased during this same time period—and emissions from fossil fuels and land use changes are one source of these emissions.

Climate remains today an extraordinarily complex area of scientific study. The risks to society and ecosystems from increases in CO_2

emissions could prove to be significant, so it is prudent to develop and implement strategies that address the risks, keeping in mind the central importance of energy to the economies of the world.[16]

Why then does ExxonMobil continue to give money to organizations that deny the statements of fact in that first paragraph?

An Industry That Cannot Afford Denial

Fossil fuel companies are in the dubious position of making more money the more carbon dioxide we emit. To find an industry to trust we need one where the opposite is true, one that would likely go broke if even the mid-case projections of global warming come true. There is such an industry: insurance. Insurance companies cannot afford to get global warming wrong; fossil fuel companies, in the short run at least, can hardly afford to get it right.

Most of us are surprised to learn of the size of the global insurance industry. Its annual revenues come to $3 trillion—three times those of Big Oil. If the revenue from all insurance companies were the gross domestic product of a single nation, that nation would be the third largest in the world. Instead of saying that the developed world runs on oil, we might be more accurate to say that it runs on insurance. Surely a modern society could not function without insurance to spread risk. And no doubt about it, global warming threatens the insurance industry's ability to function and survive. As Joel Ario, Pennsylvania Insurance Commissioner and chair of the insurers national task force on climate change put it, "The insurers are perhaps the one group that is more concerned about climate change than the environmentalists."[17]

Hurricanes, cyclones, more frequent storms in some areas and more drought in others, floods, storm surges, property damage (especially along coasts), forest fires, urban fires, crop and livestock losses, pollution-related liabilities, business interruptions, equipment breakdown, data loss, civil unrest, conflicts and even wars over food, water, refugees, immigration, and land itself—the list of potential costs of global warming to insurers goes on.

It is hard to overestimate the possible losses. Consider wildfires as one example. A study that appeared in the summer of 2009 forecast that by midcentury wildfires in the western United States would burn

54 percent more acres than at present overall, 78 percent more in the Pacific Northwest, and a whopping 175 percent more in the Rocky Mountains.[18] The study assumed only a moderate increase in temperatures of 1.6° C (3° F) by 2050, well below the worst-case trajectory, which seems to be the one we are on. Will the Rocky Mountain air still be breathable when filled with the dust and smoke particles from tens of thousands of acres of burned trees? How long will insurers provide coverage for people living near a national forest?

Corporations have their own set of risks, ones that the individual homeowner does not have to worry about. Think of the liabilities to which corrupt executives at Enron, Tyco, WorldCom, and the like subjected their companies. Corporations are at risk from executive or company negligence, breach of statutory duty, failure to take proper precautions to protect shareholder value, even breach of human rights. Liability suits could arise from materials or products that increase carbon dioxide emissions. Like Big Tobacco, companies could find themselves liable because they made a dangerous product but wove a tissue of lies to hide it. No wonder some insurance companies, notably those in Europe, have been speaking up about global warming and denying coverage where they deem the risk of weather-related disasters to be too great.

Recent history is not encouraging. As global temperatures rose over the second half of the twentieth century, U.S. insurance losses mounted, in part because people built more homes, and more expensive ones, along the Atlantic and Gulf Coasts, but also possibly in part because of global warming. Since Hurricane Andrew in 1992, eighteen national disasters costing over $1 billion each have struck the U.S. In 2004—the year *before* Katrina—hurricane damage cost the United States more than $60 billion. As insurance companies know, the losses from smaller and less publicized disasters can add up and in total be as costly as larger ones. The year 2006, seemingly a relatively benign year for weather-related disasters, saw 43 catastrophic loss events in North America and 349 globally. Losses to insurers from natural disasters in 2007 were nearly double those of 2006, due to winter storms in Europe, flooding in Britain, and wildfires in the United States. Between 1980 and 2004, global catastrophic losses worldwide totaled $1.4 trillion.

The fundamental premise of insurance is that past experience allows one to assess future risk. But global warming puts this premise itself at risk: it is already clear that if we stopped all carbon emissions *today*, the future will still be worse than the past. Legendary investor Warren

Buffett, who owns the largest American reinsurance company, said in 1992: "Our exposure goes up every year because of what's going on in the atmosphere, even though we don't fully understand what's going on. Catastrophe insurers can't simply extrapolate past experience. If there is truly 'global warming,' the odds would shift, since tiny changes in atmospheric conditions can produce momentous changes in weather patterns." The investment guru continued, "A hurricane that caused X dollars of damage twenty years ago could easily cost 10X now."[19]

To protect themselves from catastrophic losses, the familiar companies from which we buy our life and homeowner's insurance buy their own coverage from reinsurance companies. The first insurance company to express concern about global warming was giant Munich Reinsurance, all the way back in 1973. Since then, European reinsurance companies, especially Munich Re and Swiss Re, have become increasingly vocal. In 2007, at a meeting of one hundred of the world's leading insurance companies, the two joined German reinsurance company Allianz in signing an initiative calling for limits on greenhouse gas emissions. American insurers have drug their feet: of those attending the meeting, only Marsh and McLennan agreed to sign.[20] Hurricanes in 2005 cost the Bermuda-based Ace Ltd. almost $1 billion after taxes, but its chief executive was still unwilling to take global warming into account, saying, "I am agnostic."[21] According to the president of the Insurance Information Institute, "The majority of interests here [in the United States] have been either hostile or had no opinion." The cochair of a task force of the National Association of Insurance Commissioners, which regulates the U.S. industry, and who also serves as director of insurance for Nebraska, said, "We're seeing all kinds of extreme weather in the Great Plains states, including drought, tornadoes, brushfires and severe hailstorms." But, he added, "I feel like I am singing but nobody is listening. The only real industry response is 'don't tread on me.'"[22]

While American insurance companies have hesitated, the large European insurers have spoken out:

Munich Re: "Our database clearly shows that the number of weather-related natural catastrophes in Europe has more than doubled since 1980. There is increasing evidence that this trend is already driven by climate change. As a result . . . past loss experience is no longer a suitable yardstick for predicting future losses. Instead, the consequences of global warming, which vary from region to region, must be anticipated now, and reflected in pricing and risk management."[23]

Swiss Re: "Global warming is a fact. The climate has changed: visibly, tangibly, measurably. An additional increase in average global temperatures is not only possible, but very probable, while human intervention in the natural climatic system plays an important, if not decisive role. Climate change does not merely imply a possible increase in extreme levels, such as higher wind speeds or an increase in precipitation. Instead, it means above all a change in average, 'normal' weather."[24]

Allianz (the world's second-largest international insurance and financial services company, also headquartered in Munich): "Climate change poses significant risks throughout the United States, particularly to coastal, flood-prone and fire-prone areas. Without examining how global warming could intensify risk, it is impractical . . . to carry out long-term planning to protect assets. The insurance industry needs to prepare itself for the negative effects that climate change may have on its business and on its customers. On the other hand, it can significantly help mitigate the economic risks and enter the low carbon economy by providing appropriate products and services."[25]

The Association of British Insurers: "Annual losses from the three major storm types affecting insurance markets (U.S. hurricanes, Japanese typhoons and European windstorms) could increase by two-thirds to $27 bn by the 2080s. Climate change could increase wind-related insured losses from extreme U.S. hurricanes by around three-quarters to total $100–150 bn. This additional cost would be equivalent to two to three Hurricane Andrews [at that time, Andrew was the last Category 5 hurricane to make landfall in the United States] in a single season (at 2004 prices)."[26]

In 2009 the U.S. National Association of Insurance Commissioners developed a disclosure form that member insurance companies would have to fill out and return to the organization. The survey asked such questions as, "Does the company have a plan to assess, reduce or mitigate its emissions in its operations or organizations? Does the company have a climate change statement of policy with respect to risk management and investment management? If yes, please provide. If no, how do you account for climate change and risk management?" And finally, "Has the company considered the impact of climate change and global

Balance as Bias

How the Media Missed "The Story of the Century"

In a misguided attempt to be "fair and balanced," or because of a failure of understanding, outright laziness, or because controversy sells and consensus does not, American media have given the same few deniers equal weight with the world scientific community.[1] The individual deniers, organizations like the Heartland Institute and their funders—none could have succeeded in duping America had the media not aided and abetted. As expected, right-wing media like the *Wall Street Journal* and Fox News are guilty, but so are the *Washington Post* and the *New York Times*. The two-decades long success of the industry of denial could *not* have happened without the complicity of the media. How else to explain why, as the scientific evidence for global warming has risen, public acceptance has fallen?

That the media have failed is not just an opinion: it is a fact established by scholarly research. In 2004, two researchers, Maxwell T. Boykoff and Jules M. Boykoff, tested the premise that press coverage of global warming among national newspapers had misled the public.[2]

The Boycoffs concentrated on four leading national newspapers—the *New York Times*, the *Washington Post*, the *Los Angeles Times*, and the *Wall Street Journal*—searching databases such as Lexis-Nexis for the phrase "global warming" in news stories appearing between 1980 and 2002. They found 3,543 articles and from them drew a random sample of 636. Slightly over half were "balanced"—that is, they gave "roughly equal attention" to the view that humans caused global warming as they did to the opposite view: that global warming is exclusively natural. Just over one-third emphasized human-caused global warming but also discussed the possibility that global warming is natural. Slightly over 6 percent focused on the denial of global warming. Only 6 percent attributed global warming solely to human activities, thus endorsing only the consensus view of scientists.

Equally telling were the changes that the Boykoffs found over time. In 1988 the vast majority of newspaper articles emphasized scientists' emerging view of global warming. By 1990, when the first IPCC report appeared and when the Global Climate Coalition and Heartland Institute had begun their campaign of denial, reportage had shifted toward the "balanced" view. Politicians then became the group most quoted on global warming.

The Boykoffs sum up: "In the end, adherence to the norm of balanced reporting leads to informationally biased coverage of global warming. This bias, hidden behind the veil of journalistic balance, creates both discursive and real political space for the U.S. government to shirk responsibility and delay action regarding global warming.[3]

The Prestige Press

Given the much greater evidence for global warming today and the overwhelming scientific consensus, one would hope that media stories now reflect that consensus. True, right-wing media like Fox News and Rush Limbaugh have become more pervasive since the period the Boykoffs studied, but surely the "prestige press" has gotten on the right side of the issue. Read on.

On July 14, 2009, the *Washington Post* gave precious space to an opinion piece by Sarah Palin, who had just announced her resignation as governor of Alaska. The title of her article was, "The 'Cap And Tax' Dead End."[4] It was odd to find the *Post* printing an article by Palin on energy policy, for back in September 2008 the paper had said she was "way off" on energy issues. Vice presidential candidate Palin had said on more than one occasion that Alaska produces 20 percent of American domestic energy. This seemed surprising on its face, since Alaska's share of oil production has been shrinking and it has little natural gas, almost no coal, and no nuclear generators. The *Post* reported that Alaska's actual contribution is just 3.5 percent. For her sixfold error, the *Post's* ombudsman "Fact-Checker" gave Palin "Four Pinocchios"—its highest award for inaccuracy, earned only for a "Whopper."[5] This record would hardly have seemed to qualify Palin as enough of an expert on energy to entitle her to a place on the *Post*'s opinion page, yet there she was.

On December 9, 2009, the *Post* gave Palin another chance in an article titled, "Copenhagen's Political Science." Speaking of the e-mails stolen from the Climate Research Unit of the University of East Anglia, which I will cover in chapter 14, Palin said, "The e-mails reveal that leading climate 'experts' deliberately destroyed records, manipulated data to 'hide the decline' in global temperatures, and tried to silence their critics by preventing them from publishing in peer-reviewed journals." But as we now know, each of these statements is false or misleading: scientists destroyed no records, manipulated no data, and prevented no one from publishing. The "decline" referred to an anomalous set of tree-ring data that the experts replaced with actual temperature measurements.

The *Post* was not the only prestige newspaper to give space to global warming denial. In February 2009, Al Gore had addressed an overflow audience at the Chicago meeting of the American Association for the Advancement of Science. By now, he had given his "Inconvenient Truth" slideshow hundreds of times to audiences around the world. Midway through his Chicago talk, Gore brought up the wildfires that were wreaking havoc around the world—in California, Georgia, Florida, Greece, and especially in Australia—keeping up a steady narration as the slides appeared on the screen, one after another. Then Gore came to a chart that resembled Paul Bunyan's hockey stick, showing an almost vertical rise in the number of weather-related disasters: drought, extreme temperatures, floods, landslides, storm surges, wildfires, and windstorms that had occurred between 1900 and 2007. The slide, which Gore said came from the Center for Research on the Epidemiology of Disasters, showed that four times as many weather-related disasters had occurred in the last thirty years as in the fifty years before that. As he showed the slide, the former Vice President narrated: ". . . this is creating weather-related disasters that are completely unprecedented." After pausing on the Bunyan Hockey Stick for a few seconds, he moved on to the rest of his talk and dozens of additional slides.

Gore himself, and the hundreds of scientists who heard him, must have been dumbfounded a few days later to read this headline in the *New York Times*: "In Climate Debate, Exaggeration Is a Pitfall," and to find the *Times* accusing Gore of being as impervious to reason and facts as George Will, saying that *both* men are guilty of "hyperbole," "inaccuracies," "overstatement," and "hype."[6] The article identified only one Gore "exaggeration": the slide showing the rise in weather-related disasters since 1900. But the *New York Times* had itself used the very same slide in an article the previous May. In describing the chart in that article, *Times* columnist Charles M. Blow had written, "This surge in disasters and attendant costs is yet another reason we need to declare a coordinated war on climate change akin to the wars on drugs and terror. It's a matter of national security."[7]

Was Gore wrong on the facts? Two years before, a reporter summed up a meeting of insurance companies by saying that they are "acutely aware of the dramatic increase in losses caused by natural disasters in recent decades [and are] convinced that global warming is partly to blame."[8] What about the center that generated the data shown in the chart? It responded, "We believe that the increase seen in the graph until about 1995 is explained partly by better reporting of disasters in general, partly due to active data collection efforts . . . and partly due to real increases in certain types of disasters. We estimate that the data in the most recent decade present the least biased and reflect a real change in numbers. This is especially true for floods and cyclones. Whether this is due to climate change or not, we are unable to say."[9] As soon as Gore learned that he may have gone beyond what his source was willing to stand behind, he withdrew the slide from his presentation.

Neither Gore nor Will is a scientist; neither has done original scientific research; all their information is secondary. The difference is that in his public presentations and in his film and book, Al Gore presents the results of peer-reviewed scientific studies and speaks for the overwhelming scientific consensus. George Will denies that consensus and the science behind it, while presenting no counter evidence or theory of his own. Gore speaks for reason; Will for ignorance.

Plumes of Smoke from China

Global warming denial is not the only field of science to suffer from media anti-science bias disguised as balance. In 2005 the late Tim Russert, one of the most respected television journalists, decided to use *Meet the Press* to ask whether thimerosal, a mercury-bearing compound introduced in minute amounts into vaccines as a preservative, caused autism.[10] By this time, four scientific studies had exonerated thimerosal. (By 2008, four more had reached the same conclusion.) Had Russert wanted to provide his audience with the latest and best scientific information, guest Harvey Fineberg, president of the Institute of Medicine and one-time dean of the Harvard Medical School, would have said something like, "There is no scientific evidence that thimerosal causes autism in children"; then the shortest segment of *Meet the Press* in history would have ended. Sound journalism, poor television. So Russert invited a second guest: David Kirby, a journalist and author of *Evidence of Harm: Mercury in Vaccines and the Autism Epidemic: A Medical Controversy*, published in March 2005. Kirby "dismissed Fineberg's epidemiology with a wave of his hand."[11] But what may have been more important to viewers than what Fineberg or Kirby said was the mere sight of them sitting side by side in front of Russert, proof that there was a legitimate debate over the role of thimerosal. But among experts, there is no such debate. Within a few months, Kirby would go on to claim that plumes of mercury-bearing smoke from coal fires in China were causing autism in West Coast children; then he would blame California wildfires. But there was no detectable difference between the rates of autism nationally and in California children.

Peter Duesberg is the most credentialed scientist to deny that HIV causes AIDS.[12] As Richard Lindzen and others claim that dissident climate scientists cannot get research grants, Duesberg claims that the medical and HIV publishing and funding establishment has censored him. In March 2006, *Harper's* magazine did not even try for balance, publishing "Out of Control," a long article by AIDS denialist Celia Farber that presented Duesberg as a courageous dissident denied the funds necessary to prove his case.[13] Author Seth Kalichman points out that AIDS scientists found "50 significant errors and misrepresentations" in her article. In the *Columbia Journalism Review*, Gail Beckerman said that Farber, a well-known AIDS denialist, had used the article to argue "that big pharmaceutical companies have basically invented the concept of AIDS in order to sell their product, which, being extremely toxic, is what is actually killing people who are diagnosed HIV-positive."[14] In its May 2006 issue, *Harper's* published seven letters on the Farber article as well as her response. Three letters were supportive of Farber's questioning of the science of AIDS; one disagreed; two were mixed. The seventh letter was from Robert Gallo, the codiscoverer of the AIDS virus. Gallo said, "There is more evidence that HIV causes AIDS than there is for the cause of any other single human disease caused by an infectious agent, past or present." Farber dismissed Gallo's letter by saying it was "riddled with assertions of fact that dissolve under careful scrutiny into highly debatable interpretations of ambiguous data."

Carbongate

In March 2009 the Environmental Protection Agency submitted to the White House its finding that greenhouse gas emissions are pollutants that endanger public health and welfare. In June, reports began to surface that two weeks before the EPA turned in its finding, the agency had quashed an internal report that cautioned against "decisions based on a scientific hypothesis that does not appear to explain most of the available data." Were the accusation true, the damage would have been two-fold: First, the EPA would have squelched legitimate scientific dissent, revealing that it had become politicized, as the deniers have claimed for years. Second, the agency might then have based its endangerment finding on faulty science.

As evidence for its warning that global warming is overhyped, the allegedly suppressed report by EPA staff member Alan Carlin and a coauthor cited the "downward trend in temperatures since 1998 (which some think will continue until at least 2030)."[15] Global warming does not threaten more or larger hurricanes, they said; Greenland is not shedding ice; the global economic recession has lowered CO_2 emissions; the water vapor feedback is negative, not positive as the IPCC assumed; solar activity could account for up to 68 percent of observed global warming. Charts and tables filled the report, giving the impression that it had come from EPA climate scientists.

Fox News swallowed the report whole, saying "Global cooling. Who knew?" The Manchester (New Hampshire) Union Leader titled its editorial, "Warm Skepticism; Jury Out on Climate Change." "The EPA silences a climate skeptic," gushed the Wall Street Journal. The Copley News Service circulated a piece titled, "It's Getting Cold Out There." And finally—as the astute reader would have predicted—a new term appeared in the media: "Carbongate." The media took the bait.

To the deniers in Congress, the flap could not have come at a better time—just before members were to consider the Waxman-Markey climate and energy bill. Senator Inhofe protested, "Over the last few days . . . we have learned that a senior EPA official suppressed a detailed, rigorous account of the most up-to-date science of climate change."[16] Joe Barton was in high dudgeon: "The science is not there to back [the EPA finding] up," he said. "An EPA report that has been suppressed . . . raises grave doubts about the endangerment finding."[17] Puffed Rep. James Sensenbrenner (R-WI), "I'm sure it was very inconvenient for the EPA to consider a study that contradicted the findings it wanted to reach. But the EPA is supposed to reach its findings based on evidence, not on political goals. The repression of this important study casts doubts on EPA's finding, and frankly, on other analysis EPA has conducted on climate issues."[18]

But the chart did not come from scientists. Its lead author, Dr. Alan Carlin, is an economist. More importantly, as anyone could have found out from an Internet search, whole sections of the report appear to have been lifted nearly verbatim from a denier website, Patrick Michaels's World Climate Report.[19] Compare these two selections (TSD refers to the EPA's Technical Support Document):

World Climate Report Website	Carlin Report
. . . the Endangerment TSD is largely a dated document which relies heavily on the *Fourth Assessment Report* (AR4) of the U.N.'s Intergovernmental Panel on Climate Change (IPCC). The IPCC's *AR4* was published in the spring of 2007, but to meet the deadline for inclusion in the *AR4*, scientific papers had to be published by late 2005/early 2006. So, in the rapidly evolving field of climate change, by grounding its TSD in the IPCC *AR4* the EPA is largely relying on scientific findings that are, by late 2008, nearly 3 years out of date.	The draft endangerment TSD is largely a dated document which relies heavily on the Fourth Assessment Report (AR4) of the U.N.'s Intergovernmental Panel on Climate Change. The IPCC's AR4 was published in the spring of 2007, but to meet the deadline for inclusion in the AR4, scientific papers had to be published by late 2005/early 2006. So, in the rapidly evolving field of climate change, by grounding its TSD in the IPCC AR4 the EPA is largely relying on scientific findings that are, by early 2009, largely three years or more out of date.

The EPA responded, "Claims that [Carlin's] opinions were not considered or studied are entirely false. [He] is not a scientist and was not part of the working group dealing with this issue. Nevertheless the document he submitted was reviewed by his peers and agency scientists, and information from that report was submitted by his manager to those responsible for developing the proposed endangerment finding. In fact, some ideas from that document are included and addressed in the endangerment finding."[20]

But even a report of dubious provenance might still be right. As reflected in the comparison above, the report's principal argument is that because the IPCC's AR4 is out of date, policymakers cannot rely on it as a basis for action. It is true: the IPCC's Fourth Assessment is out of date. But it is out of date in the wrong direction for the deniers: as has been shown repeatedly, AR4 *underestimated* the speed and the extent of global warming, such that the IPCC's worst-case scenario looks increasingly likely not to be bad enough. If the deniers want to replace AR4 with an accurate, up-to-date analysis, they are going to be disappointed at its findings.

The lesson of "Carbongate" is how easily the media and deniers in Congress swallow denier claims, making little or no effort to uncover their true sources and their motivations. No matter how transparent and false, thanks to the media such claims take on a life of their own. Repeatedly, deniers latch on to some new finding that, they say, finally falsifies global warming and exposes the conspiracy. Then the new fatal flaw turns out to be nothing of the sort; things quiet down until deniers discover the next fatal flaw, then the process repeats, and the media and we fall for it again. Meanwhile, carbon dioxide concentrations and temperatures rise and precious time is lost.

End of Objectivity?

Why do the print media continue to give equal time to the deniers? We may find a clue in a change that took place in the Code of Ethics of the Society of Professional Journalists. For years the code had included a section titled "accuracy and objectivity." But after a 1996 revision, "objectivity" disappeared and the code read, "Public enlightenment is the forerunner of justice and the foundation of democracy. The duty of the journalist is to further those ends by seeking truth and providing a fair and comprehensive account of events and issues."[21]

To most of us, being "fair and comprehensive" means presenting both sides of a debate. When newspapers report on politics, such balance is essential. After all, if we know anything, it is that in politics there are no permanently right answers. Today's minority opinion about tax policy, education, healthcare, welfare—you name it—has a good chance of becoming tomorrow's majority opinion. In political reporting, good journalism requires presenting the minority point of view, so that the reader can understand the debate. But science is different. In science, there are right answers and wrong answers. Over time, with fits and starts, science moves steadily toward the right answers—toward the truth. For example, scientists no longer debate whether continents drift. For a reporter to present both sides of that now settled argument—and no doubt any diligent reporter could find at least one contrarian geologist who would claim that continents do not move—would do readers a disservice. Granted, it might sell more newspapers. Unless there is a substantial debate among credentialed, practicing scientists, to present a fair and balanced account about a matter of science is to present the consensus view.

Harvard's Eric Pooley describes the three stances that a reporter can take.[22] One is to be a *stenographer*: recording the give-and-take of debate without taking sides or giving any indication which side is apt to be correct. At best, stenography adds little or no value; at worst, it misleads the reader.

At the other extreme, pundits play *judge and jury*, deciding for the reader what is right. Instead of providing the evidence so that we readers can make up our own minds, they make them up for us.

Pooley recommends that reporters adopt the third stance, that of *referee*, keeping the game honest, calling fouls where they occur and telling the reader who committed the foul. A referee would call a technical foul and eventually eject from the game the for-hire deniers, who are like athletes who take money under the table to shave points.

On an issue as vital as global warming, the public needs to know the truth, and it cannot unless the media make it possible. Unless there is a demonstrated lack of consensus, the media should not attempt to provide two sides on a question of science, but present the consensus view. Otherwise the media not only mislead, they endanger the public.

American journalism has a chance to get it right, but so far it continues to fail the test.

Systemic Failure

On September 4, 2009, the *Washington Post* headlined an article: "Emissions Linked to End of 2,000-Year Arctic Trend."[23] The source was a report in *Science* showing that Arctic temperatures were now warmer than at any time in the last two millennia. The conclusion came from the work of "30 researchers from the United States, Britain, Denmark, Norway, Canada and Finland [who had] reconstructed the Arctic's climate in the distant past." After quoting some of those scientists, the *Post* article continued, "Fred Singer, a prominent climate-change skeptic, questioned the *Science* study, saying it does not properly reflect other researchers' findings about the Medieval Warm Period. That period, between A.D. 800 and 1300, had 'higher temperatures than even the past 30 years,' he said." The article went on to question Singer's statement, but why include it at all? Why quote Singer in the first place? Especially when he is dead wrong. Why does he, who has done no climate research in decades, deserve an equal place with today's practicing climate scientists?

Next let us turn to the paper of record, the *New York Times*. The following is a list of the headlines of articles by *Times* reporters on global warming that appeared in the first five months of 2010.

"U.N. Panel's Glacier Warning Is Criticized as Exaggerated" (January 19)
"Past Decade Was Warmest Ever, NASA Finds" (January 22)
"Less Water Vapor Slows Earth's Warming Trends, Researchers Say" (January 29)
"Researcher on Climate Is Cleared in Inquiry" (February 4)
"Skeptics Find Fault with U.N. Climate Panel" (February 8)
"Climate Change Debate Is Heating Up in Deep Freeze" (February 11)
"U.N. Climate Chief Quits, Deepening Sense of Disarray" (February 18)
"Scientists Taking Steps to Defend Work on Climate" (March 3)
"Darwin Foes Add Warming to Targets" (March 4)
"Panel Will Review U.N. Climate Work" (March 11)
"Among Weathercasters, Doubt on Warming" (March 28)
"Climate Fears Turn to Doubts Among Britons" (May 24)

As far as one can tell from the headlines, ten of the twelve articles are about the "controversy," not about the science of global warming. Two articles do appear to be about science. One of those says that the earth's warming trend appears to have slowed. By the end of January, when that article appeared, the first decade of the twenty-first century was known to be the warmest on record (as an article the week before had headlined), 2009 was known to have tied for the second warmest year globally, and in the Southern Hemisphere, to have been the warmest. The earth's warming trend had not slowed.

The headline of the other possible science-based story, "Past Decade Was the Warmest Ever . . . " seems to get the science right, but what about the rest of the article? After noting that NASA found the decade ending in 2009 to have been the warmest on record, the reporter, John Broder, adds that the "new temperature figures are not apt to be the last word on whether the planet's temperature is on a consistent upward path." Why tack on that qualifying phrase? Because, Broder writes, "[NASA's] Dr. Hansen, who has been an outspoken figure in the climate debate for years, has often been attacked by skeptics of global warming for what they charge is selective use of temperature data. The question of whether the planet is heating and how quickly was at the heart of the so-called 'Climategate' controversy that arose last fall when hundreds of e-mail messages from the climate study unit at the University of East Anglia in England were released without authorization."

But none of the Climategate e-mails, which were stolen, not "released," gave any hint that Hansen had been at fault. Moreover, even though all NASA's temperature data and methods are open to the public, no critic has ever shown that Hansen used data selected to exaggerate warming. There is simply not one iota of doubt that the first decade of this century was the warmest on record. But anyone reading Broder's article would have concluded that, to the contrary, there is plenty of room for doubt because Hansen and other scientists may have cheated.

Any open-minded person, reading those twelve headlines and Broder's full article, would be bound to come away with increased uncertainty about global warming, even though just in the five months during which the stories appeared, the actual evidence had grown stronger. The overall effect of the twelve stories is to mislead readers about the most important scientific issue of the day. American media have

Science Under Attack

On SkepticalScience.com, John Cook lists and in one short sentence each refutes more than 130 denier arguments against global warming.[1] The site is also available as a mobile application, allowing ready access during a conversation around the family dinner table or at your favorite watering hole. I have already covered some denier claims; in this chapter I will examine several other prominent ones.

It's the Sun, Ozone, Volcanoes, Sulfates

That global warming is natural, not man-made, is one of the earliest and most obvious arguments. Before humans began to add carbon dioxide to the atmosphere, all changes in climate had to have been natural. We know from the geologic record that both CO_2 and temperature have varied in the past, so we cannot dismiss the argument that today's changes might also be natural. In addition to variation in the Sun's output, the other possible sources of natural climate variability are atmospheric ozone and volcanic eruptions, which inject carbon dioxide, dust, and sulfur into the atmosphere.

Ozone, a molecular form of oxygen with three atoms, absorbs ultraviolet rays in the upper atmosphere. Thus less atmospheric ozone would mean more ultraviolet radiation reaching the surface, warming the earth and providing a natural explanation for observed twentieth-century global warming. Atmospheric ozone concentrations did decrease from

1980 until about 1995, but as adherence to the Montreal Protocol began to reduce CFCs (chlorofluorocarbons), ozone started to increase. Ozone levels and global temperatures do not appear to correlate; evidently the effect is too small to matter.

Volcanoes emit carbon dioxide, but only 300 million tons annually, just 1 percent of human-caused emissions. They also emit sulfates, which form tiny droplets, or aerosols, that reflect solar rays and thus tend to cool the earth, as happened after the eruption of Mount Pinatubo in 1991. But these aerosols wash out on a scale of months, whereas CO_2 remains in the atmosphere and oceans for centuries.

Variations in the output of the Sun appear to be the only natural cause that might possibly explain global warming. Not only does the Sun supply virtually all of Earth's surface energy, the Sun's activity and global temperatures have correlated in the past. From 1880 until about 1975, solar activity and global temperatures tracked closely. But then the correlation stopped: solar activity has stayed about the same during the period in which global temperatures rose sharply. The year 2009 had the lowest solar activity in fifty years but was the second warmest year on record.

One way to examine possible causes of global warming other than greenhouse gases is to use a single climate model to test the effect of each cause one at a time.[2] For example, we know that as humans burned more coal during the twentieth century, the amount of atmospheric sulfur rose. Sulfur in the atmosphere forms tiny particles that reflect sunlight and have an overall cooling effect. The Parallel Climate Model (PCM) of the U.S. Department of Energy shows that had sulfate been the only factor at work, instead of warming during the twentieth century the earth would have cooled by about 0.25°C (0.45°F).

The PCM also allows an estimate of the effect on global temperatures of the observed variation in the intensity of the Sun's rays. Had the Sun been the only factor, temperatures would have risen about 0.2°C (0.4°F). Modelers repeated the experiment for each of the other three factors—volcanic eruptions, ozone, and greenhouse gas emissions—as though each was the only factor in play. Summing the effects of solar, ozone, volcanic, and sulfate, they find that the net effect on temperature is close to zero: these factors together do not come close to explaining the observed rise in global temperature. But when the modelers add greenhouse gases to the mix, the modeled temperatures closely match those actually observed.

These results come from the use of only a single climate model, whereas scientists today use some two dozen models and average the results. The experiment single-handedly debunks two denier arguments: that something other than greenhouse gas emissions can explain the temperature record of the twentieth century; and that climate models do not work. I will have more to say about climate models at the end of this chapter.

Testosterone

Another denier argument is that the amount of carbon dioxide from fossil fuels is too small to make a difference. It is true that when we speak of units of "parts per million" it seems we are talking about tiny amounts. But since 1800, carbon dioxide concentrations have increased from 280 to 390 ppm, or by nearly 40 percent. Moreover, the increase has come on top of a natural system that is in balance. The oceans and land plants and animals emit about 780 gigatons (billion tons) of carbon annually, and absorb nearly all of it. Human activities emit 29 gigatons of carbon per year, but since we humans absorb none of our emissions, that carbon enters the atmosphere as CO_2, where the greenhouse effect causes temperature to rise. But are we sure that the extra carbon in the atmosphere actually came from fossil fuels? Yes: studies of carbon isotopes prove it.

Carbon has three naturally occurring isotopes, each with six protons (which is what makes them carbon), but different numbers of neutrons. Carbon-12 has six neutrons, carbon-13 seven, and carbon-14 eight. During photosynthesis, plants have a predilection for the lightest isotope, carbon-12, absorbing relatively more of it and winding up with a lower ratio of carbon-13 to carbon-12 (C-13/C-12) than the atmosphere. When plants die and change into coal, oil, and natural gas, they retain their original isotopic ratios. The total amount of a carbon-based substance in one's body, such as testosterone, can increase or decrease for different reasons. But as cyclist Floyd Landis found out the hard way, synthetic and natural testosterone have different carbon isotope ratios and they point to the source as natural or man-made. People may lie, but isotopes do not.

Since plants have a lower ratio of carbon-13 to carbon-12, as we burn more fossil fuels, made from decomposed plants, the relative amount of carbon-13 in the atmosphere should fall. At the Mauna Loa Station, scientists measured not only the amount of carbon dioxide in the atmosphere but also its isotopic ratios. As the amount of atmospheric carbon dioxide rose over the decades, the ratio of carbon-13 to carbon-12 fell steadily, showing that more of the carbon had come from the burning of fossil fuels. There is no other explanation than that humans are responsible for the increase in atmospheric carbon dioxide concentrations.

Scientists also measure carbon isotope ratios in air bubbles trapped in ancient ice cores. They find that as soon as the amount of carbon dioxide in the atmosphere began to rise with the Industrial Revolution, the isotope ratio began to fall and kept falling. At no time in the last 10,000 years was the ratio of carbon-13 to carbon-12 as low as it is today.

Watts Up with That?

If deniers could show that temperatures are *not* rising, or even that we cannot be sure they are, they would have made their case. To that end, deniers make a set of temperature-related claims. For instance, they allege that modern temperature measurements exaggerate global warming because they include the "urban heat island effect," as Michael Crichton did in *State of Fear*. We know that modern cities, with their tall buildings and acres of steel and concrete, are warmer than the surrounding countryside. Thus increasing urbanization alone might make it appear that the planet is warming. But scientists filter out the urban data and, when they do, find that though slightly diminished, the warming trend remains. The IPCC said in its Fourth Assessment, "Studies that have looked at hemispheric and global scales conclude that any urban-related trend is an order of magnitude [factor of ten] smaller than decadal and longer time-scale trends."[3] Thus the urban heat island effect turns out to be small and easily corrected.

In a related argument, deniers say that the U.S. historical temperature record is unreliable because some weather stations are sited near trees, buildings, parking lots, air conditioners, and the like, causing the stations to record unrepresentative, and presumably warmer, local temperatures. The person most behind this claim is Anthony Watts, a former weatherman at radio station KPAY in Chico, California. In 2007, Watts founded SurfaceStations.org to demonstrate that "some of the global warming increase is not from CO_2 but from localized changes in the temperature-measurement environment."[4] By early June 2009, thanks to a grass roots network of volunteers, SurfaceStations.org had examined about 70 percent of the 1,221 stations in the National Oceanographic and Atmospheric Administration's Historical Climatology Network, enough to find out if there is anything to Watts's claim. SurfaceStations.org classified 70 of the stations as having "good or best" reliability—that is, being least affected by local environmental factors. If those 70 stations showed a different temperature trend than the other 1,151, the deniers would have a point. But since 1950, temperatures recorded at the 70 "good or best" stations are indistinguishable from the rest. Here is how NOAA summed up:

> We would expect some differences simply due to the different area covered: the 70 stations only covered 43% of the country with no stations in, for example, New Mexico, Kansas, Nebraska, Iowa, Illinois, Ohio, West Virginia, Kentucky, Tennessee or North Carolina. Yet the two time series . . . are remarkably similar. Clearly there is no indication from this analysis that poor station exposure has imparted a bias in the U.S. temperature trends.[5]

Slap Shot

The most persistent accusation of mistaken temperature measurements focuses on the "hockey stick" chart of global temperatures, first presented by Michael Mann, Raymond Bradley, and Malcolm Hughes. Figure 12.1 shows the 2001 version as presented in the IPCC's Third Annual Report of that year.

The gray error bars before, say, 1600 are wide enough that one cannot have much confidence in the location of the white line of averages prior to that date. Perhaps in part for that reason, critics of the hockey stick immediately spoke up.

The most vocal were Stephen McIntrye and Ross McKitrick, who published three articles on the hockey stick. Two appeared in *Energy and Environment*, a non-peer-reviewed journal. They published the third in the highly respected and peer-reviewed *Geophysical Research Letters*.[6] It argued two points: First, that Mann, Bradley, and Hughes had used an "unusual data transformation" that can conjure up a hockey stick out of nothing but noise. Second, they had relied too much on tree-ring

Northern Hemisphere

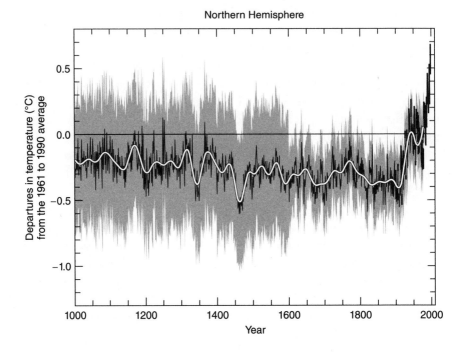

Figure 12.1 The hockey stick, 2001. Black = individual measurements; gray = error bars; white = average.

data from two species: Bristlecone and Foxtail pines. For both reasons, McIntrye and McKitrick claimed, the apparent hockey stick shape was not statistically significant.

The second of the two articles in *Energy and Environment* went further, accusing Mann, Bradley, and Hughes of scientific misconduct and citing a number of examples "where results adverse to their claims were not reported (and in some cases, actual misrepresentations)."[7]

Given its iconic status, it is no surprise that the hockey stick controversy reached the press, the public, and Congress. At the time, Rep. Joe Barton chaired the House Energy and Commerce Committee while Rep. Ed Whitfield (R-KY), chaired the Subcommittee on Oversight and Investigation. The two wrote to Mann, Bradley, and Hughes demanding their résumés, a list of all financial support for their research, any agreements they had made to disseminate and share their results, the locations of all data archives, their "exact computer code," a list of all requests for data that Mann, Bradley, and Hughes had received and their reply, their response to the errors alleged by McIntrye and McKitrick, and a detailed explanation of their work for and on behalf of the IPCC's Third Assessment Report, including the identity of "the people who wrote and reviewed the historical temperature-record portions of the report particularly Section 2.3, 'Is the Recent Warming Unusual?'"[8] Barton set up his own review panel, prompting prominent science supporter Rep. Sherwood Boehlert (R-NY), now retired, to ask the National Academy of Sciences to review the hockey stick. Then ensued the "Battle of the Panels."

Heading the Barton panel of three statisticians was Edward Wegman, professor at George Mason University. The Wegman Report found Mann, Bradley, and Hughes "to be somewhat obscure and incomplete and the criticisms . . . to be valid and compelling." Moreover, "Mann, Bradley, and Hughes's claim that the decade of the 1990s was the hottest decade of the millennium and that 1998 was the hottest year of the millennium cannot be supported by [Mann's] analysis."[9] Wegman said in oral testimony, "Method wrong plus answer correct is just bad science," as though Mann, Bradley, and Hughes had made a wild guess and gotten lucky.[10] Accusing Mann, Bradley, and Hughes of "bad science" and saying the criticisms were "compelling" was bound to leave the impression that the hockey stick was wrong. But Wegman never said that.

Twelve experts from fields including climate science, geochemistry, statistics, and meteorology comprised the Academy panel that reviewed the hockey stick paper. The panel confirmed the original findings, writing, "The committee finds it plausible that the Northern Hemisphere was warmer during the last few decades of the twentieth century than during any comparable period over the preceding millennium."[11] The panel noted that the criticized statistical techniques had made a difference of only about 5/100 of a degree to the final numbers. Moreover, removing the Bristlecone and Foxtail pine data made essentially no difference to the overall shape of the hockey stick.

The chair of the Academy panel, Gerald R. North, Distinguished Professor of Meteorology and Oceanography at Texas A&M, did not buy Wegman's assessment that the Mann, Bradley, and Hughes paper was "bad science." In an interview with the *Chronicle of Higher Education*, North said, "There is a long history of making an inference from data using pretty crude methods and coming up with the right answer. Most of the great discoveries have been made this way. The Mann et al. results were not 'wrong' and the science was not 'bad.' They simply made choices in their analysis which were not precisely the ones we (in hindsight) might have made. It turns out that their choices led them to essentially the right answer."[12]

The accusation led the Mann, Bradley, and Hughes team, as well as other scholars, to test and improve the hockey stick. Two researchers published a review of the original hockey stick in 2007.[13] They concluded that "a slight modification to the original Mann et al. reconstruction is justifiable for the first half of the 15th century, which leaves entirely unaltered the primary conclusion of Mann et al. (as well as many other reconstructions) that both the 20th century upward trend and high late-20th century hemispheric surface temperatures are anomalous."

In 2008, Mann, Bradley, and Hughes published in the *Proceedings of the National Academy of Sciences* a new and improved version of the hockey stick that left no doubt as to its validity.[14] As John Cook of Skeptical Science puts it, "the science of paleoclimatology has moved on." Scientists have found hockey stick–like patterns in borehole temperatures; in stalagmites; and in temperatures inferred from the retreat of glaciers.

The Medieval Warm Period

A related argument goes something like this: "It was warmer during the Medieval Warm Period than today; therefore today's warming is not unusual and may well be natural." The Medieval Warm Period lasted roughly from about 950 to 1250 C.E., allowing the Vikings to sail an ice-free sea and establish colonies on Greenland and North America. Until recently, estimates of temperatures during the Medieval Warm Period were for the entire globe, but in 2009 Michael Mann and colleagues published an article in *Science* that used proxies (tree rings, corals, etc.) to reconstruct temperatures during the period in various regions.[15] They found that temperatures during the Medieval Warm Period were highly variable, some regions having even higher temperatures than today, others having lower ones, in some cases even below temperatures during the Little Ice Age that followed the Medieval Warm Period. The North Atlantic, parts of Greenland, the Arctic, and North America were warmer during the Medieval Warm Period; Central Asia, northwestern North America, and the central Pacific were cooler. The effects were regional, not global, and overall the Medieval Warm Period was not as warm as today.

Both global warming deniers and climate scientists ought to be relieved that temperatures during the Medieval Warm Period were not as high as today's. Suppose for the sake of argument that they had been— that during the Medieval Warm Period, Nature had achieved today's higher temperatures *without* benefit of human-caused CO_2 emissions. Then imagine the temperature rise that would occur if we had the equivalent of a Medieval Warm Period due to natural causes and, on top of it, the additional global warming that must result from CO_2 concentrations over 100 ppm higher. Be careful what you wish for.

Do Climate Models Work?

Can we rely on the climate models, our only guides to future climate? The models must prove themselves and that will not be easy because Earth's climate is one of the most complex systems scientists have ever tried to model. The first computer models of climate were primitive by today's standards, yet even they showed that Arrhenius with his pencil and paper had gotten close to the right answer for how much temperature would rise if carbon dioxide concentrations were to double. Today's models run on supercomputers and include the atmosphere, the oceans, sea ice, the biosphere, and all known feedbacks. As we saw earlier in this chapter, when modelers add the effects of greenhouse gases to the set of natural causes of global temperature change, the models replicate twentieth-century temperatures quite well.

Climate models have been around long enough to have already made several successful predictions, the ultimate test of their validity. In 1988, James Hansen used a climate model to project future temperatures using three different paths that carbon dioxide emissions might take in the future. Today we know that Hansen's business-as-usual projection was right on the money, needing adjustment only because he did not know that Mount Pinatubo would erupt in 1991.

Climate models have made several other accurate projections:

- As the surface of the earth warms, the stratosphere should cool. It has.
- The layer of the atmosphere right above the surface, the troposphere, should warm. At first, temperatures measured from satellites contra-

dicted this prediction. But the satellite calculations from the University of Alabama at Huntsville turned out to be in error. After correction, the observations are in line with model projections.

• In 1992, after the Pinatubo eruption but before its effects had been measured, Hansen's model predicted almost exactly what those effects turned out to be.[16]

• The models predict that ocean surface waters will warm and they have. Sea surface temperatures during the summer of 2009 were the warmest on record.

• Climate feedbacks should amplify global warming in the Arctic; temperatures there have risen more than at lower latitudes.[17]

These successes aside, given the complexity of the problem, computer models can provide us only with a range of future possibilities. It will be a long time, if ever, before they can predict the effects of global warming, especially precipitation, on a regional scale. But they are the best guide we have.

Did Global Warming End in 1998?

Figure 12.2 (from NASA) shows how, while rising and falling in the short term, global temperatures have risen over the long term and especially since about 1970, with 2010 and 2005 tied for the hottest year on record. Those who argue that there has been no global warming since 1998 base their claim instead on the temperature record from Britain's Hadley Center, which uses a slightly different approach and which had shown 1998 as the hottest calendar year on record. Scientists have a good idea of why 1998 was a high temperature year. The greatest cause of year-to-year temperature variations is the El Niño–La Niña Southern Oscillation cycle. During El Niño years, the oceans give up heat to the land, causing land surface temperatures to rise; during La Niña years, the opposite effect occurs. The year 1998 saw the "El Niño of the century"; 2008 was a moderate La Niña year. So one would expect 1998 to have been anomalously hot and 2008 to have been anomalously cool, as figure 12.2 shows they were. To gauge whether global temperature is rising over the long run requires a longer period than ten years.

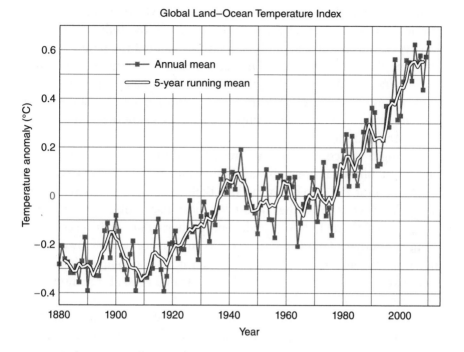

Figure 12.2 Global land-ocean temperature anomalies relative to a 1951–1980 base period. The point for 2010 is preliminary and through November (NASA).

Chicken or Egg?

"There is no greenhouse effect—temperature drives carbon dioxide increases, not the other way round," says the Heartland Institute's *Skeptic's Handbook*. We cannot dismiss this argument out of hand, for temperature and carbon dioxide are so closely linked that in trying to decide which leads which, we confront a chicken versus the egg dilemma. But we can resolve it by following temperature and carbon dioxide as the earth cycles in and out of an ice age. Only a few decades ago, this would have been an impossible dream, but the remarkable ice cores have allowed it.

In polar regions, new layers of snow continually bury the old; the rising pressure with depth gradually converts the snow into ice. The key to understanding past climates is that falling snow traps atmospheric dust from deserts, ash from volcanic eruptions, and bubbles of air. Surprising as it may seem, the ice retains the air bubbles, which are like fossils of the ancient atmosphere at the time the snow fell. By studying the ratios of oxygen isotopes in the bubbles, scientists can tell the temperature at the time the bubble formed. They also measure the amount of carbon dioxide in the bubbles, as though a time machine had transported them back scores of thousands of years. The ice cores give information about ancient forest fires, meteorite impacts, volcanic eruptions, atmospheric chemistry, and more.

During the Second Soviet Antarctic Expedition in 1957, Russian scientists established a research station that they named Vostock (Russian for "East") for the ship of Fabian von Bellingshausen, an Antarctic pioneer. The Soviets chose the coldest and most inaccessible spot on Earth, near the Southern Pole of Cold, the Southern Pole of Inaccessibility, and the Southern Geomagnetic Pole. Vostock recorded the lowest temperature ever measured: −128.6° F. The air is almost completely dry and lacks oxygen and carbon dioxide. The polar night lasts three months of the year. Conditions are ideal!

Starting in the 1970s, the Russians began to drill down into the Antarctic ice and extract long cores. Joined later by the French, they obtained a core over 3,600 meters (2.2 miles) long, which stretched back 420,000 years. Because the Vostock holes were approaching a pristine sub-ice lake, the drillers halted at that depth; a European consortium called EPICA began new drilling 500 kilometers (km) away.

The EPICA cores extend the record back over 800,000 years, revealing eight ice age cycles during which temperature, carbon dioxide, and methane varied identically. That the cycles repeated with virtual metronomic precision shows that the cause of the ice ages lay not in the earth, but in the stars. The cores beautifully captured the astronomical cycles identified by Serbian scientist Milutin Milankovitch, thus answering the ancient question that had eluded Arrhenius, Callendar, and many others: what caused the Ice Ages. Now we know.

The ice cores allow such precise measurements that scientists can tell that as the earth emerges from an ice age, temperature rises first, then on average about 800 years later, carbon dioxide begins to rise. After that, the two rise together in an amplifying feedback loop. The deniers have made much of the initial temperature rise, though scientists have explained repeatedly that the astronomical cycles cause it.

Another remarkable fact, and a cautionary one, is how small are the changes that Milankovitch discovered. The dominant effect is the

100,000-year cycle in the earth's eccentricity: as Earth orbits the Sun, the shape of its path changes from slightly more elliptical to slightly more circular and back again. The pattern causes a tiny change in the amount of sunlight striking the Southern Hemisphere, coincident with spring warming there. The warming begins to melt Antarctic sea ice and Southern Hemisphere glaciers. As the oceans warm slightly, they are unable to hold as much carbon dioxide and release a small amount, which builds up in the atmosphere. Then the greenhouse effect takes over to amplify the tiny temperature increase far beyond what the change in eccentricity alone could have caused. Wallace Broecker said of the earth's climate system, "If you're living with an angry beast, you shouldn't poke it with a sharp stick."[18] A tiny jab can produce a disproportionate response.

As was the case with the hockey stick and other disputed evidence, when scientists find a discrepant or disputed result, they go back to work to try to discover the reason. The rise of temperature and carbon dioxide in the ice cores is a superb example, telling us several things. First, that the Milankovitch astronomical cycles are real. Second, that they initiate the warm, interglacial periods and are the main driver of the ice ages. Third, that temperature and carbon dioxide are so intimately linked that if one rises, the other also rises. Fourth, climate change can be extremely rapid. During some glacial-interglacial transitions, temperatures rose by several degrees, not in a matter of centuries or even decades, but in years. Deniers claim that because temperature starts to rise a few centuries before carbon dioxide, the greenhouse effect and global warming must be wrong. Instead, the fact turns out to be a source of vital information about Earth's climate and a testament to the ingenuity and fecundity of a great scientific theory. And to the hardiness of Antarctic scientists.

GRACE

On a warming Earth, more ice would melt and polar ice sheets would shrink. Yet according to the deniers, Antarctic ice is growing. They are right, but it's the Antarctic *sea* ice that is growing. The *land* ice is shrinking dramatically.

Antarctica covers 5.5 million square miles—how do scientists know how much ice and snow blanket the continent? Until recently, the answer would have been: they don't. Then in 2002, NASA and the German Aerospace Center launched a satellite to conduct the Gravity Recovery and Climate Experiment: GRACE. By measuring the pull of gravity as it sails far overhead, the GRACE satellite is able to calculate the mass of Antarctic ice. Since 2002, Antarctica has lost more than 100 cubic km (24 cubic miles) of ice per year.

But since 1979, the sea ice around the margins of the huge continent has grown, suggesting that temperatures there have cooled. Instead, air temperature in the Southern Ocean has been warming. How could sea ice near Antarctica be expanding at the same time air temperatures are rising? Because the amount of sea ice depends not only on air temperature, but on water temperature. The complete story is complicated, but part of the answer is that ocean currents are carrying less heat to the surface to melt the ice that forms near West Antarctica.[19]

Tropospheric Cooling?

As noted above, climate models predict that the troposphere, the layer closest to the earth's surface, should warm faster than the surface. The models also predict that the stratosphere, the zone from the top of the troposphere at 4.3 miles up to 11 miles, should cool, and it has. But in the early 1990s, satellite measurements indicated that since 1979 the troposphere had been cooling, not warming, casting doubt on the climate models and perhaps on global warming itself.[20]

The interpretation of satellite temperature data is complicated. Scientists cannot use the data directly, but must apply several corrections. Weather satellites have no internal power sources, so over time their orbits decay, requiring scientists to apply a correction. It turns out that the scientists responsible for the original satellite measurements had applied the wrong numerical sign to that correction.

When others pointed out the error and scientists made proper corrections, the troposphere turned out to have been warming at about the same rate as the surface, as the climate models predicted. Temperatures for the troposphere in the Tropics still appear anomalous, so as usual, answering one question raises others, a frustrating but ultimately enlivening characteristic of science.

One could go on rebutting each of the remaining denier arguments—and John Cook has. But the pattern is the same: the deniers scavenge the work of legitimate scientists to find any discrepant result and, when they discover one, boast that global warming "was never a crisis." Scientists find new evidence that removes the discrepancy, or discover that it resulted from an error, which they correct, only to have the deniers go right on repeating the same discredited claim.

As of this writing, there are no research findings that falsify global warming. The deniers have no uncontested facts on their side and no counter theory. That is why, as a last, intellectually bankrupt ploy, they claim that scientists have made up global warming.

Greatest Hoax in History?

In a speech on the Senate floor on July 28, 2003, Oklahoma Sen. James Inhofe called the threat of catastrophic global warming "the greatest hoax ever perpetrated on the American people." Two years later, far from recanting, Inhofe hearkened back to that speech with evident pride, adding that "environmental extremists exploit the issue for fundraising purposes, raking in millions of dollars, even using federal taxpayer dollars to finance their campaigns."[1] By early 2009, Inhofe had lost his position as chair of the Senate Committee on Environment and Public Works to Californian Barbara Boxer. In another speech on January 8, Inhofe did not use the word "hoax," but his opinion had not changed: "For the last six years, I have been talking about the Hollywood and media-driven fear . . . that tries to convince us that those who are fueling this machine called America are somehow evil and fully responsible for global warming."[2]

Inhofe led a "truth squad" to the Copenhagen climate conference in December 2009, where a question produced this exchange:[3]

REPORTER: "If there's a hoax, then who's putting on this hoax, and what's the motive?"
INHOFE: "It started in the United Nations and the ones in the United States who really grab ahold of this is the Hollywood elite."

After fielding a few more questions, the one-man truth squad headed to the airport.

Stephen Moore is a *Wall Street Journal* editorial board member, Cato Institute senior fellow, *National Review* contributing editor, and regular CNBC and Fox News commentator. In an address to "New Jersey Citizen Activists" in May 2009, Moore said to loud applause: "I happen to believe that global warming is the biggest scam of the last two decades."[4] Harold Ambler, writing on *Huffington Post*: "Mr. Gore has stated, regarding climate change, that 'the science is in.' Well, he is absolutely right about that, except for one tiny thing. It is the biggest whopper ever sold to the public in the history of humankind."[5] Bob Lutz was in charge of product development at General Motors, including its plug-in hybrid car, the Volt. His opinion of global warming? It's "a total crock . . . "[6] John Coleman, founder of the Weather Channel and meteorologist at KUSI in San Diego, wrote:

> [Global warming] is the greatest scam in history. I am amazed, appalled and highly offended by it. Global Warming; It is a SCAM.
> In time, a decade or two, the outrageous scam will be obvious. As the temperature rises, polar ice cap melting, coastal flooding and super storm pattern all fail to occur as predicted everyone will come to realize we have been duped.[7]

Who's to Blame? Liberals, of Course

As noted earlier, in his keynote address at the 2009 Heartland Institute Conference, meteorologist Richard Lindzen dismissed the opinions of one of his MIT colleagues about global warming because his "politics are clearly liberal." Rush Limbaugh heads his website, "Despite Cooling Temperatures, Liberals Still Sell Global Warming." A book published in 2008 has the title, *The Really Inconvenient Truths: Seven Environmental Catastrophes Liberals Don't Want You to Know About—Because They Helped Cause Them*. Recall that Canadian denier Dr. Timothy Ball said that global warming is "the political agenda of a group of people . . . who believe that industrialization and development and capitalism and the Western way is a terrible system and they want to bring it down."[8] In a paper reprising his 2009 Heartland Institute Conference talk, Harrison Schmitt wrote, "Given what we actually know about climate, and all the remaining uncertainties, Americans should think long and hard before giving up more of their liberties and income to satisfy politicians who just want to 'do something' to satisfy a particular special interest. A long-term political agenda is at work here, gathering power at the expense of liberty."[9]

If the deniers can blame scientists' claims about global warming on their allegedly left-leaning politics, then the debate becomes just one more polarized topic in our polarized nation, just one more reason for the public to reserve judgment, or to tune out altogether, and for policymakers to stand pat.

To Command Spring

What would it take for Inhofe and Moore and the other deniers who believe global warming is a con trick to be right? First, scientists would have to be so devoted to an ideological cause that they would risk their careers. But most scientists are much more interested in research than politics.

Second, scientists would have had to fabricate the raw data that provides the evidence for global warming. They would have had to spend years, even entire careers, engaged in a criminal conspiracy to fake tens of thousands of data points and thousands of scientific papers.

Moreover, scientists would have had to command spring to arrive earlier and fall to show up later; glaciers to retreat; Greenland and Antarctica to melt faster each year; Arctic sea ice volume to decline; snow cover to shrink; permafrost to thaw; sea level to rise; oceans to acidify; birds to migrate; plants to flower earlier in the spring; Northern Hemisphere species to migrate north and upslope; and western wildfires to increase in frequency, size, and duration.

Third, the scientific hoaxsters would have to be able to get away with the fraud. But American universities and government agencies have elaborate procedures for detecting scientific misconduct. The more eminent a researcher and the more important the claims, the harder others will try to replicate them and, if they fail, the sooner and louder they will blow the whistle.

Harvard professor of psychology Marc Hauser reported in 1995 that cotton-top tamarind monkeys could recognize themselves in mirrors, an ability that up to then scientists thought only humans and great apes possessed. Researcher Gordon Gallup found the claim hard to believe and requested Hauser's film of the experiments, reporting that he saw "no resemblance whatever to the reported data."[10] In 2001, Hauser retracted the claim. Though it would have been better for him to have found his own mistake, scientists and funding agencies are apt to forgive a researcher who publicly owns up to an error. Nevertheless, the admission may arouse suspicions.

On August 10, 2010, the *Boston Globe* reported that Hauser had been under investigation for alleged misconduct since 2007.[11] As a result, he had retracted a 2002 study and published a correction to one from 2007, both of which asserted that the cognitive abilities of monkeys were more like those of humans than scientists believed. On August 27, 2010, Harvard announced that it had put Hauser on leave and that he may have fabricated data in the 2002 paper.[12] If authorities substantiate the charge, agencies will likely ban Hauser from receiving research grants and he may lose his position at Harvard.

Though the Hauser case is still being adjudicated, we know that other scientists have cheated and otherwise bent the rules. The same is true of every profession known to man. But scientists who cheat are caught, reprimanded or banned, and science marches on. Science is far better than scientists.

Fourth, in order to keep the hoax a secret, nearly all scientists would actually have to *be* liberal, however one defines the term, and be in on the plot. A survey by the Pew Foundation found that 52 percent of American scientists identify themselves as liberal, leaving 48 percent to blow the whistle.[13] But global warming is a global subject. Roughly half of IPCC scientists are from other countries, where the terms *liberal* and *conservative*, on which we Americans obsess, have a different meaning or no meaning. Liberal American scientists would have had to dupe their foreign colleagues into going along with their deceit in order to foster the Americans' political agenda. It is unclear why scientists from Brazil, China, India, and Russia, say, would be so dumb and so accommodating as to risk their professional standing for the sake of our internecine political squabbles.

If it is difficult to get away with charging scientists with being fraudulent conspirators, governments, which history shows often are corrupt, make a more plausible target. In a 2001 denial manifesto in the *Wall Street Journal*, Lindzen, who served on the IPCC, wrote that, "The [IPCC] Summary for Policymakers . . . represents a consensus of government representatives (many of whom are also their nations' Kyoto representatives), rather than of scientists. The resulting document has a strong tendency to disguise uncertainty, and conjures up some scary scenarios for which there is no evidence."[14]

Does that accusation pass the smell test? Remember that governments created the IPCC to moderate scientists by ensuring that government bureaucrats vet the scientists' reports. The comparison (cited in chapter 2) of the scientists' draft and the final version of the 2007 IPCC Summary for Policymakers showed dozens of instances where the bureaucrats had watered down the scientists' conclusions. Not once did government representatives strengthen the scientists' statement. After the IPCC report came out, non-IPCC scientists roundly criticized it—not for being too strong, but for being too weak. If the government bureaucrats tried to strengthen and dramatize the IPCC reports, as the deniers claim, they failed miserably.

Is it possible that the IPCC representatives of the Bush administration such as James Connaughton, chair of the Bush White House's Council on Environmental Quality and former lobbyist for power companies, asbestos industries, and large electricity users, is more "alarmist" on global warming than the average IPCC scientist and succeeded in exaggerating evidence for global warming? The same James Connaughton whose office pressured the EPA to dramatically weaken its public statements about New York air quality in the days after the 9/11 attacks?

Fifth, if the world's scientists wanted to bring down "industrialization and development and capitalism and the Western way," why did they wait to use global warming as the vehicle? Every energy company employs scientists. Years, decades ago, geologists working for Exxon-Mobil could have fudged their maps and reports just enough to lead the company to drill in the wrong places and eventually go bankrupt. Of course, those geologists would have lost their jobs, even their careers. Hurricane scientists could have kept their projected storm trajectories to themselves. Computer specialists could have nudged their weather forecast programs just enough to hide oncoming extreme weather events. The opportunities boggle the mind. The notion might make an excellent plot for a science fiction novel.

Funding Research

Another denier refrain is that scientists create alarm over global warming to keep themselves in research funds. Lindzen says that his MIT departmental colleague Carl Wunsch "assiduously avoids association with skeptics; if nothing else, he has several major oceanographic programs to worry about." Another MIT colleague, Kerry Emanuel, "received relatively little recognition until he suggested that hurricanes might become stronger in a warmer world. He then was inundated with professional recognition." (How one would like to be a fly on the wall at an MIT Meteorology Department cocktail party!) Lindzen reserves special opprobrium for Columbia's Wallace Broecker, "who staunchly beats the drums for alarm and is richly rewarded for doing so."[15]

Modern scientific research requires money, no doubt about that, and finding it is a constant worry and duty of any research scientist. Writing grant proposals, administering them, and reporting to funding agencies occupy a large fraction of the time of any productive research scientist. That is the price they pay for doing what they like to do: research.

The federal government is by far the major funder of scientific research in the United States, the National Institutes of Health and the National Science Foundation being the two largest supporters. The process works like this: Researchers submit proposals to a funding agency; experts outside government review and rank the proposals; agencies fund those that receive the highest rankings, as budgets permit. The process has a natural bias toward mainstream science, which agencies try to resist by "sunsetting" some grant areas and by allocating a percentage of funding for bold and potentially breakthrough new ideas. Overall, the process has worked remarkably well, otherwise American science would not be the best in the world.

Overseeing the National Science Foundation is a National Science Board, a group appointed by the President, each of whose members swears to protect and defend the Constitution. I served on the National Science Board for twelve years, appointed first by President Reagan and then by President George H. W. Bush. Some members of the NSB were liberal; some were deeply conservative. (The first fellow NSB member I met was William Nierenberg, who also served on the boards of Fred Singer's Science and Environmental Policy Project (SEPP) and the George C. Marshall Institute, and who signed the Leipzig Declaration. At the time—1986—like most scientists, I would have had no idea what the two organizations and the declaration stood for, instead preferring to quietly mind my own business.) I can attest that during the twelve years I served, the political persuasions of NSB members had no influence on NSF policy, though they did make for interesting cocktail party chatter.

No doubt some scientists have chosen to work on climate issues, rather than on some other topic closer to their true interests, because they believe funding will be available. Without funding, most scientists cannot do research. Presumably most of us have chosen careers in which we think we can at least make a living and a contribution. Because a scientist chooses to do research in an area where funding is available, does it follow that the scientist would fake research results to get the next grant? Even if large numbers of scientists are that dishonest, it would not work. NIH and NSF peer-review, conducted by experts who know as much about the proposed research as the proposal writer, would quickly spot such fabrications. An agency would debar the proposal writer from ever receiving another government research grant.

Another thing is wrong with the claim that scientists raise alarm about global warming merely to secure their next research grant: during the second Bush administration, global warming scientists employed by the federal government were more apt to lose research funding—or even their jobs—than gain it. After Hansen spoke at the 2005 American Geophysical Union's meeting about the dangers of global warming, his bosses at NASA ordered public relations staff to vet his public statements. Later, NASA's inspector general found that in attempting to squelch Hansen, NASA had acted improperly. Scientists at the National Oceanographic and Atmospheric Administration had similar experiences and feared for their jobs. Can there be any doubt that federal scientists at earlier stages in their career than Hansen, with less distinguished records, got the message?

The effort to squelch climate scientists was part of an attack on science under George W. Bush, as well reported by Chris Mooney in his fine book, *The Republican War on Science*. Scientists finally had enough and fired back. On February 18, 2004, over 62 leading scientists—Nobel laureates, medical experts, former federal agency directors, and university chairs and presidents–signed the statement below on scientific integrity prepared by the Union of Concerned Scientists. The statement read in part:

> When scientific knowledge has been found to be in conflict with its political goals, the [Bush] administration has often manipulated the process . . . by placing people who are professionally unqualified or who have clear conflicts of interest in official posts and on scientific advisory committees by disbanding existing advisory committees, by

censoring and suppressing reports by the government's own scientists, and by simply not seeking independent scientific advice. The administration has sometimes misrepresented scientific knowledge and misled the public.[16]

Traitors

For global warming to be a hoax, climate scientists, research laboratories, university administrators, funding agencies, scientific journals, the United Nations, government science agencies, ministers, and diplomats would all have to be part of a vast international conspiracy unprecedented in human history. The hoaxsters would have had to create a foolproof mechanism to keep whistleblowers and loudmouths quiet, lest they spill the beans. Science would have to be a corrupt criminal enterprise to dwarf the Mafia.

But surely, the reader must be thinking, the notion that global warming is a hoax is absurd on its face. Surely, no one but a demented crackpot could truly believe that. Read on.

On Friday, June 26, 2009, the U.S. House of Representatives took up the first major climate bill to have a chance of passing the Congress: the American Clean Energy and Security Act (ACES), known as the Waxman-Markey bill for its two sponsors. When it came his turn to speak, Rep. Paul Broun (R-GA), a medical doctor in his former life, said, "Scientists all over this world say that the idea of human-induced global climate change is one of the greatest hoaxes perpetrated out of the scientific community. It is a hoax. There is no scientific consensus."[17] After Broun's remarks, his Republican colleagues gave him a hearty round of applause.

Have deniers like Broun thought through the implications of their claim that global warming is a hoax? If it is true, then those who accept, or pretend to accept, that global warming is real and that the United States needs to act now to prevent it are misleading and endangering the nation. One might almost accuse them of treason. It is curious then, that some of the most distinguished members of the intelligence and defense communities, not to mention President Obama and many members of Congress, are among the potential traitors.

On June 25, 2008, Dr. Thomas Fingar, Deputy Director of National Intelligence for Analysis and Chairman of the National Intelligence

Council, testified before the House Select Committee on Energy Independence and Global Warming. The title of his testimony was "National Intelligence Assessment on the National Security Implications of Global Climate Change to 2030."[18]

Dr. Fingar said that for information on climate science, the U.S. intelligence community relied on the "Intergovernmental Panel on Climate Change (IPCC) Fourth Assessment Report, which we augmented with other peer-reviewed analyses and contracted research. We used the UN Panel report as our baseline because this document was reviewed and coordinated on by the U.S. government and internationally respected by the scientific community." He reported that

> Climate change could threaten domestic stability in some [nations], potentially contributing to intra- or, less likely, interstate conflict, particularly over access to increasingly scarce water resources. . . . From a national security perspective, climate change has the potential to affect lives (for example, through food and water shortages, increased health problems including the spread of disease, and increased potential for conflict), property (for example through ground subsidence, flooding, coastal erosion, and extreme weather events).

And those were the concerns that he said would arise by 2030. What would happen after that date, Dr. Fingar did not speculate.

What about the military? Are they part of the conspiracy of traitors? Since sitting military officers may be unable to speak freely, let us see what former high-ranking U.S. military personnel have to say. In 2007 the Military Advisory Board, which provides information to the U.S. Navy and comprises eleven retired generals and admirals, issued a report called *National Security and the Threat of Climate Change*.[19] Each of the eleven high-ranking officers signed his name by hand to the report. Its introduction read in part:

> Global climate change presents a new and very different type of national security challenge.

> Carbon dioxide levels in the atmosphere are greater now than at any time in the past 650,000 years, and average global temperature has continued a steady rise. This rise presents the prospect of significant

climate change, and while uncertainty exists and debate continues regarding the science and future extent of projected climate changes, the trends are clear.

The U.S. military has a clear obligation to determine the potential impacts of climate change on its ability to execute its missions in support of national security objectives.

The decision to act should be made soon in order to plan prudently for the nation's security. The increasing risks from climate change should be addressed now because they will almost certainly get worse if we delay.

The Military Advisory Board report included personal statements from the retired officers. "We will pay for this one way or another," wrote Gen. Anthony C. Zinni, a Marine and former head of Central Command. "We will pay to reduce greenhouse gas emissions today, and we'll have to take an economic hit of some kind. Or we will pay the price later in military terms. That will involve human lives," General Zinni warned. "It's not hard," he wrote, "to make the connection between climate change and instability, or climate change and terrorism."

General Zinni's medals include the Bronze Star, Purple Heart, Vietnam Service Medal (with three service stars), Combat Action Ribbon, Navy Distinguished Service Medal, and the Defense Distinguished Service Medal (with one oak leaf cluster).

But the conspiracy may reach higher in the military establishment than General Zinni. On February 1, 2010, Secretary of Defense Robert M. Gates submitted to Congress his department's Quadrennial Defense Review. President George H. W. Bush had nominated Gates as Director of the Central Intelligence Agency, and in November 1991 the Senate confirmed him. He was serving as president of Texas A&M University when President George W. Bush tapped him in 2006 to succeed Donald Rumsfeld as Secretary of Defense. The Armed Services Committee unanimously confirmed Gates; the full Senate approved by a vote of 95 to 2. Gates agreed to stay on as Secretary of Defense under President Obama.

Under Gates's signature, the 2010 Quadrennial Defense Review noted that

Climate change and energy are two key issues that will play a significant role in shaping the future security environment. Climate-related changes are already being observed in every region of the world. Among these physical changes are increases in heavy downpours, rising temperature and sea level, rapidly retreating glaciers, thawing permafrost, lengthening growing seasons, lengthening ice-free seasons in the oceans and on lakes and rivers, earlier snowmelt, and alterations in river flows. . . . Climate change could have significant geopolitical impacts around the world, contributing to poverty, environmental degradation, and the further weakening of fragile governments. Climate change will contribute to food and water scarcity, will increase the spread of disease, and may spur or exacerbate mass migration.[20]

On the issue of global warming, who are the true patriots?

Climategate

Much Ado About Nothing

After the release of the IPCC's Fourth Assessment Report in the spring of 2007, international climate conferences under the auspices of the United Nations took place in Bali the following December and in Poznan, Poland, in December 2008. These laid the groundwork for the Copenhagen conference in mid-December 2009, whose goal was to replace the Kyoto Protocol, set to expire in 2012, with steeper, mandatory reductions in carbon emissions. As the UK's *Guardian* put it in an editorial, "The politicians in Copenhagen have the power to shape history's judgment on this generation: one that saw a challenge and rose to it, or one so stupid that we saw calamity coming but did nothing to avert it. We implore them to make the right choice."[1]

On November 19, 2009, websites began to buzz with news that an anonymous hacker had stolen a trove of e-mails from the Climate Research Unit (CRU) of the University of East Anglia (UEA), along with NASA one of the two main repositories and presenters of global temperature information. Soon the entire set of more than 1,000 stolen e-mails appeared online in a searchable database, and hopes for the Copenhagen conference crashed.[2] Denier websites and columnists exulted in the news that the e-mails exposed global warming as a hoax, just as they had always known it to be. No one seemed interested in the actual theft, as though burglary is fine as long as it uncovers a clandestine and prurient diary.

James Delingpole of Britain's *Daily Telegraph* got out his purple pen: "The conspiracy behind the Anthropogenic Global Warming myth has been suddenly, brutally and quite deliciously exposed."[3] FoxNews.com answered its own question affirmatively: "Do E-mails Reveal Scientist Claims on Climate Change are . . . BUNK?" Patrick Michaels (profiled in chapter 7) opined in the *Wall Street Journal* in a piece titled, "How to Manufacture a Climate Consensus." The subhead read, "The East Anglia emails are just the tip of the iceberg. I should know." He accused the authors of the stolen e-mails of "silencing climate scientists" and said the revelations "have dramatically weakened the case for emissions reductions."[4] The *Telegraph* quoted Sen. James Inhofe on what the politician said was "the worst scientific scandal of our generation."[5] No sooner had the story appeared than someone dubbed the episode "Climategate." Scientists named in the e-mails began to receive threats, including death threats, prompting the British police and the FBI to investigate.

Nearly all of the stolen e-mails were routine exchanges among scientists, but a few contained lines that at first glance appeared to lend credence to the deniers' accusations. The most cited may be this one from Phil Jones, director of the Climate Research Unit: "I've just completed Mike's [Michael Mann of hockey stick fame] trick of adding in the real temps to each series for the last 20 years (i.e. from 1981 onward) and from 1961 for Keith's to hide the decline." The language led Sarah Palin to claim in her op-ed in the *Washington Post* of December 9, 2009, that "leading climate 'experts' . . . manipulated data to 'hide the decline' in global temperatures," and later to refer to "snake-oil" science. But that is not what Jones's e-mail shows.

Recall that to measure temperatures before people recorded them with thermometers and satellites—before about 1880—scientists have to use proxies: temperatures determined from tree rings, coral reefs, cave deposits, ice cores, and the like. Obviously, these proxies are not as precise as a thermometer, hence the wide error bars on the original hockey stick; but if you have enough proxy measurements, they verify each other and give reliable results. Scientists had found and discussed in the literature that, after 1960, temperatures estimated from tree-rings from certain far northern trees show a decline, even though thermometer measurements show temperatures were rising. Mann's "trick" was to use in his reconstruction the actual temperatures since the 1960s, not the inexplicable tree-ring proxy data. Thus the "decline" was not in global temperatures, but in a set of obviously incorrect tree-ring data. Mann's "trick" made the result more reliable, not less.

Palin, Michaels, and other deniers claim that the e-mails show that scientists "tried to silence their critics by preventing them from publishing in peer-reviewed journals," as Palin put it. This accusation arises from an e-mail from Mann complaining about one published article and saying, "Perhaps we should encourage our colleagues in the climate research community to no longer submit to, or cite papers in, this journal." The journal in question was *Climate Research* and the article was the one by Sallie Baliunas and Willie Soon that I discussed in chapter 9, in which the authors claimed "that the 20th century is probably not the warmest nor a uniquely extreme climatic period of the last millennium." The American Petroleum Institute funded the Baliunas-Soon study. After the article's publication, three of the journal's editors, including the incoming editor in chief, resigned in protest. Thirteen scientists whom Baliunas and Soon cited in the article published a rebuttal. Mann was not trying to silence critics, but was instead suggesting that his colleagues avoid a journal with the evident low standards of *Climate Research*.

Another charge had to do with the articles reviewed by the IPCC as it prepares its periodic reports. On July 8, 2004, Jones wrote to Mann: "The other paper by MM is just garbage. . . . I can't see either of these papers being in the next IPCC report. Kevin and I will keep them out somehow—even if we have to redefine what the peer-review literature is!" ("MM" refers to Stephen McIntrye and Ross McKitrick, who published articles attempting to discredit the hockey stick, as discussed in chapter 12.) Rajendra Pachauri, chairman of the IPCC, says that the two papers "were actually discussed in detail in chapter six of the Working Group I report. Furthermore, articles from the journal *Climate Research*, which was also decried in the emails, have been cited 47 times in the Working Group I report."[6] Thus Jones's unfortunate e-mail had no effect, neither silencing critics nor preventing anyone from publishing in peer-reviewed journals.

In another e-mail to Mann, Jones suggests deleting a data file rather than allowing anyone to have access. He also recommends that Mann delete e-mails he may have exchanged with a member of the Climate Research Unit staff and asks Mann to urge others to do likewise. These suggestions were unwise at best. But the larger point is that according to Trevor Davies, Pro-Vice Chancellor of Research at the university, no files were deleted or "otherwise dealt with in any fashion with the intent of preventing the disclosure." "Leading climate 'experts'" did not "deliberately destroy records."

Next we come to an e-mail from Kevin Trenberth, head of the Climate Analysis Section at the National Center for Atmospheric Research in Boulder, Colorado. Trenberth wrote:

> We are not close to balancing the energy budget. The fact that we can not account for what is happening in the climate system makes any consideration of geoengineering quite hopeless as we will never be able to tell if it is successful or not! It is a travesty! The fact is that we can't account for the lack of warming at the moment and it is a travesty that we can't.

Deniers used this language to claim that a leading scientist hid his lack of confidence in the climate models and in global warming itself. Delingpole wrote that the words show Trenberth "[c]oncealing private doubts about whether the world is really heating up."[7]

In the rest of the e-mail Trenberth refers to an article he wrote called "An Imperative for Climate Change Planning: Tracking Earth's Global Energy," in which he says, "Given that global warming is unequivocally happening and there has so far been a failure to outline, let alone implement, global plans to mitigate the warming, then adapting to the climate change is an imperative."[8] Far from admitting doubt, Trenberth describes global warming as unequivocal. The "travesty" is that scientists do not have better measuring systems and better understanding of the earth's energy balance.

One Million Words

With the e-mails available online, scientific organizations and responsible media read them and drew their own conclusions. Five staff members of the Associated Press perused each of 1,073 e-mails, about 1 million words in total, and concluded that, "E-mails stolen from climate scientists show they stonewalled skeptics and discussed hiding data—but the messages don't support claims that the science of global warming was faked."

Here is a sample of other opinions on Climategate:

- The American Geophysical Union: the emails were "being exploited to distort the scientific debate about the urgent issue of climate change."[9]
- The American Association for the Advancement of Science reaffirmed its conclusion that "based on multiple lines of scientific evidence global climate change caused by human activities is now underway, and it is a growing threat to society." Alan I. Leshner, CEO of the AAAS and executive publisher of *Science*: "It's important to remember that the reality of climate change is based on a century of robust and well-validated science."[10]
- *Nature*: "To these denialists, the scientists' scathing remarks qualify as the proverbial 'smoking gun': proof that mainstream climate researchers have systematically conspired to suppress evidence contradicting their doctrine that humans are warming the globe. This paranoid interpretation would be laughable were it not for the fact that obstructionist politicians in the US Senate will probably use it next year as an excuse to stiffen their opposition to the country's much needed climate bill. Nothing in the e-mails undermines the scientific case that global warming is real—or that human activities are almost certainly the cause. That case is supported by multiple, robust lines of evidence, including several that are completely independent of the climate reconstructions debated in the e-mails."[11]
- Union of Concerned Scientists: "We should keep in mind that our understanding of climate science is based not on private correspondence, but on the rigorous accumulation, testing and synthesis of knowledge often represented in the dry and factual prose of peer-reviewed literature."[12]

Innocent of All Charges

In response to the publication of the stolen e-mails and the resulting slew of accusations, several investigations began. None were the result of specific, formal charges against any of the scientists named in the stolen e-mails. Rather they stemmed from the desire of universities to make sure that no one had violated their internal procedures.

The Pennsylvania State University, where Michael Mann is a professor, had begun to receive numerous e-mails, letters, and phone calls accusing Mann of manipulating data, destroying records, and colluding to squelch scientific opinion. Even though no one brought a formal charge, because "the accusations, when placed in an academic context, could be construed as allegations of research misconduct," the university decided to investigate under its Research Misconduct Policy. The most relevant section of that policy is that research misconduct can include "fabrication, falsification, plagiarism or other practices that seriously deviate from accepted practices within the academic community for proposing, conducting, or reporting research or other scholarly activities." The investigation began on November 24, 2009, and the panel reported out on February 3, 2010, finding that there exists *"no credible evidence* [emphasis mine]" that Michael Mann:[13]

> Had or has ever engaged in, or participated in, directly or indirectly, any actions with an intent to suppress or to falsify data.

> Had ever engaged in, or participated in, directly or indirectly, any actions with intent to delete, conceal or otherwise destroy emails, information and/or data related to AR4, as suggested by Dr. Phil Jones. Dr. Mann has stated that he did not delete emails in response to Dr. Jones' request. Further, Dr. Mann produced upon request a full archive of his emails in and around the time of the preparation of AR4.

> Had ever engaged in, or participated in, directly or indirectly, any misuse of privileged or confidential information available to him in his capacity as an academic scholar.

The panel said the further question of whether Mann might have "deviated from accepted practices" in climate science was beyond its purview, recommending that a separate committee of Mann's peers take up that question. On June 4, 2010, an investigative committee of the university concluded unanimously that "there is no substance to the allegation against Dr. Michael E. Mann. Dr. Mann did not engage in, nor did he participate in, directly or indirectly, any actions that seriously deviated from accepted practices within the academic community for proposing, conducting, or reporting research, or other scholarly activities."[14]

On March 31, 2010, the report of another investigation of the Climategate e-mails appeared, this one from the Science and Technology

Committee of England's House of Commons, convened because "The disclosure of climate data from the Climatic Research Unit (CRU) at the University of East Anglia (UEA) in November 2009 had the potential to damage the reputation of the climate science and the scientists involved."[15]

The report spoke plainly:

Professor Jones's actions were in line with common practice in the climate science community.

We are content that the phrases such as "trick" or "hiding the decline" were colloquial terms used in private e-mails and the balance of evidence is that they were not part of a systematic attempt to mislead. Likewise the evidence that we have seen does not suggest that Professor Jones was trying to subvert the peer review process. Academics should not be criticised for making informal comments on academic papers.

That said, the report did criticize the university for its "failure to grasp fully the potential damage by the non-disclosure of Freedom of Information Act requests."

Another panel, established by the University of East Anglia in consultation with the Royal Society, reported out on April 12, 2010.[16] The chair was Lord Oxburgh, former head of the earth sciences department at Cambridge, former chief scientific adviser to Britain's defense ministry, and former chairman of Shell. Other members included scientists from the United Kingdom and the United States. The report made it clear that, "The Panel was not concerned with the question of whether the conclusions of the published research were correct. Rather it was asked to come to a view on the integrity of the Unit's research and whether as far as could be determined the conclusions represented an honest and scientifically justified interpretation of the data."

The Oxburgh Panel concluded that it "saw no evidence of any deliberate scientific malpractice in any of the work of the Climatic Research Unit and had it been there we believe that it is likely that we would have detected it. Rather we found a small group of dedicated if slightly disorganised researchers who were ill-prepared for being the focus of public attention." The panel did find it "surprising that research in an area that depends so heavily on statistical methods has not been car-ried out in close collaboration with professional statisticians," though it also said it was unclear whether "better methods would have produced significantly different results." On April 19, after press stories on the report had appeared, the panel issued an addendum: "Neither the panel report nor the press briefing intended to imply that any research group in the field of climate change had been deliberately misleading in any of their analyses or intentionally exaggerated their findings. Rather, the aim was to draw attention to the complexity of statistics in this field, and the need to use the best possible methods."

In early July 2010, another panel convened by the University of East Anglia and chaired by Sir Muir Russell issued its 160-page report.[17] The panel's charge was to examine the behavior of the CRU scientists, "such as their handling and release of data, their approach to peer review, and their role in the public presentation of results." The Muir Russell panel reported that

- The rigour and integrity [of the CRU scientists] are not in doubt.
- We did not find any evidence of behavior that might undermine the conclusions of the IPCC assessments.
- There has been a consistent pattern of failing to display the proper degree of openness, both on the part of the CRU scientists and on the part of UEA.
- The CRU was not in a position to withhold access to temperature data or to tamper with it. We find no evidence of bias [in data selection].
- On the allegations that there was subversion of the peer review or editorial process we find no evidence to substantiate this.
- The overall implication of the allegations was to cast doubt on the extent to which CRU's work could be trusted and should be relied upon and we find no evidence to support that implication.

In September 2010, another report appeared, this one presented by the British government in response to the March report from the House of Commons Science and Technology Committee. While endorsing earlier suggestions as to how scientists can improve their procedures and communication, the report concluded that, "The focus on Professor Jones and CRU has been largely misplaced. His actions were in line with common practice." As to accusations of dishonesty stemming from the "hide the decline" language, the report found "there is no case to answer."[18]

Climategate has so far engendered six separate, independent reports by distinguished panels. Have any other allegations in the history of science been this thoroughly investigated? Yet the net effect of the panels is to reveal not a single faked data point, not a single deleted e-mail, not a single article prevented from publication. Despite all the hoopla, Climategate has made not one whit of difference to the evidence for global warming. What the e-mails do provide is the decisive test of whether climate scientists are conspiring to fake the evidence for global warming. If they are, then such key players as Jones, Mann, and Trenberth would have to be in on the hoax. Yet more than one thousand e-mails, comprising more than one million words, reveal not the slightest evidence of conspiracy.

The stolen e-mails instead reveal scientists who are human and who get frustrated, just like the rest of us. Scientists who said things they should not have said, just like the rest of us. Let he who is without sin cast the first stone (and post on the Internet his last 1,000 e-mails).

Before moving on to other "gates," let us note that no one has ever brought a formal charge of "fabrication, falsification, or plagiarism" against a single climate scientist.

Gates and More Gates

Before Climategate came "Carbongate" (the allegedly squelched EPA report described in chapter 11); after Climategate came "Glaciergate." In its Fourth Assessment, the IPCC wrote (sourcing the World Wildlife Fund):[19]

> Glaciers in the Himalaya are receding faster than in any other part of the world (see Table 10.9) and, if the present rate continues, the likelihood of them disappearing by the year 2035 and perhaps sooner is very high if the Earth keeps warming at the current rate. Its total area will likely shrink from the present 500,000 to 100,000 km2 by the year 2035 (WWF, 2005).

A literal interpretation of the first sentence is that 100 percent of Himalayan glaciers are likely to disappear by 2035, and possibly even sooner. One does not have to be a glaciologist to recognize this as a dubious claim.[20] Could all that ice really melt that fast? The second sentence is ungrammatical and contradicts the first by putting the likely shrinkage by 2035 at 80 percent. The blogosphere and the media erupted with this evidence that the IPCC had made a mistake, and a rather ridiculous one at that. The IPCC admitted that the statement was wrong, improperly sourced, and should not have been included. The error does not obscure the fact that as global warming continues, Himalayan glaciers will continue to melt. Indeed, the world's glaciers have lost mass steadily since 1980, and in the last ten to fifteen years, at an accelerating pace. The IPCC, and any human enterprise, can make mistakes. Nature does not make mistakes.

Next came "Amazongate": in another chapter of its Fourth Assessment Report, the IPCC wrote that

> Up to 40% of the Amazonian forests could react drastically to even a slight reduction in precipitation; this means that the tropical vegetation, hydrology and climate system in South America could change very rapidly to another steady state, not necessarily producing gradual changes between the current and the future situation.[21]

The citation for the statement was a report by the WWF—the World Wildlife Fund—rather than to a peer-reviewed article. Such allegedly improper sourcing incensed the denier blogosphere and led Delingpole to write, "AGW [anthropogenic global warming] theory is toast. So's Dr Rajendra Pachauri [head of the IPCC]. So's the Stern Review. So's the credibility of the IPCC."[22] Some claimed the WWF report had not even included the 40 percent figure. But a search of the article showed that it had. Moreover, as George Monbiot pointed out in the *Guardian*, the 40 percent figure came from scientific articles referenced elsewhere in the IPCC report, not from the WWF.[23] The IPCC had gotten the 40 percent and its meaning right, but used the wrong citation.

In response to the various "gates" and the criticism of the IPCC, the United Nations and the IPCC asked the InterAcademy Council, a multinational organization of science academies, to review the IPCC's processes and procedures. At the end of August 2010, the council made a prepublication copy of the report available online.[24] Chaired by Harold Shapiro, a former president of Princeton University, the InterAcademy panel concluded that "the IPCC assessment process has been successful overall and has served society well." Nevertheless, the report stated, the IPCC needs to improve its procedures by, for example, modernizing its management structure, strengthening the review process, and being more specific and informative about the degree of uncertainty in its projections. Headlines in response to the report were mostly predictable, Fox News writing, for example, "Independent Audit Panel Slams U.N.'s Climate Group."[25] The *Economist*, in its online edition, ended its article this way: "The IPCC is a unique and remarkable institution; the governments that make it up will soon have the opportunity to improve it, if they can agree about just how much reform they want, and who they want to lead it."

Let us stipulate that the IPCC has made mistakes. It needs to try even harder. But no matter how the IPCC changes its procedures, in reports of thousands of pages errors are apt to continue to creep in. IPCC members are not faceless, infallible automatons: they are overworked human beings, like the rest of us. The deniers will continue to pounce on any future errors and supply new "gates"—alleged scandals of the century that turn out not to amount to a hill of beans. But no "gate" will be able to show that the IPCC is wrong about global warming. Glaciers will continue to melt; rising temperatures will continue to threaten rain forests and the indigenous peoples and species that depend on them. To paraphrase the parting remark attributed to Galileo at the end of his trial by the Roman Inquisition: *Eppure si riscalda*. And yet it warms.

Anatomy of Denial

Global warming deniers use tactics familiar not only because the tobacco companies used them on us or our parents, but because many are rhetorical devices known since the Greeks.

Adopt the trappings of your opponent. Organizations like the Heartland Institute put on conferences with all the earmarks of scientific meetings. Introduced as experts, deniers present talks, show charts and tables, and take questions from the audience. But such conferences are Potemkin villages of denial, creating a façade of science in its absence.

Crichton's *State of Fear* is a masterpiece of illusion, a fictional polemic disguised as a work of science through the inclusion of charts, quotations, and a bibliography, all backed up by "three years" of research. Crichton fooled the American Association of Petroleum Geologists, whose spokesman said that although *State of Fear* is fiction, "it has the absolute ring of truth."

Accuse your opponent of the very thing of which you yourself are guilty. Crichton was also the master of this technique. After Sen. James Inhofe made *State of Fear* required reading for his committee, after Crichton met for an hour privately with President George W. Bush, the author repeatedly said that it was *scientists* who have politicized the issue of global warming.

The Red Herring is an argument that may be valid but, even if it is, has no bearing on the question. It aims to distract, to change the subject. The name may have come from English fox hunting, in which hunters drag a dried, smoked, and red-colored herring across the trail to throw the hounds off the scent and prolong the chase. A good example comes from the declaration the Heartland Institute issued after its 2008 conference, which said "carbon dioxide is . . . a necessity for all life." It is, but too much carbon dioxide will heat the planet to the danger point.

Deniers frequently point out, as do our friends and neighbors, that some local, temporary weather condition contradicts global warming. A midsummer cool spell brings amused smiles at the foolishness of those pointy-headed scientists who only recently claimed the earth is warming. A blizzard befalls Washington, D.C., and Sen. James Inhofe and family build an igloo and invite Al Gore to take up residence therein. But of course, global warming is about climate, not short-term weather.

Congressman Pierre S. "Pete" du Pont, former Delaware governor and Republican presidential candidate, in an article in the *Wall Street Journal*'s OpinionJournal.com, came up with a textbook example of the Red Herring: "Mars is warming significantly. NASA reported last September that the red planet's south polar ice cap has been shrinking for six years. As far as we know few Martians drive SUVs or heat their homes with coal, so its ice caps are being melted by the sun—just as our Earth's are."[1] Du Pont turned the Red Planet into a Red Herring.

In the *Straw Man* approach, one attributes to an opponent a position as easily defeated as a dummy made of straw. The opponent may not even hold the position; or if he does, it may be so trivial as to make no difference to the argument. "I will believe in evolution when a monkey gives birth to a human baby!"

Du Pont, in the same article, provides another example. In a *Washington Post* column on March 8, 2006, author David Ignatius had written that "human activity is accelerating dangerous changes in the world's climate." Du Pont responded by saying that "it is not clear that human activity is wholly responsible" for global warming. Ignatius had not said global warming was "wholly responsible" and scientists do not claim that it is.

Appeal to Authority. Find an expert witness who will lend credentials to your case. This is the tactic the denier organizations use when they enlist scientists like S. Fred Singer to argue against global warming. But Singer has also claimed that CFCs do not cause dangerous ozone depletion, that secondhand smoke is not a health hazard, and so on.

Cherry-picking. Select only the data that support your case. In his 1988 testimony before Congress, Hansen showed three projections of future temperature increases, based on three different assumptions of CO_2 emissions. As noted in chapter 8, in his testimony ten years later,

Patrick Michaels used only Hansen's most extreme projection, leaving out the other two, to allege that Hansen erred by "300 percent." In *State of Fear*, Crichton repeated Michaels's claim, giving the falsehood a wide audience. Or take the choice of 1998 as the year when "global warming ended." Pick 1997, and the claim falls apart.

The Blizzard. In his *Vanity Fair* interview (discussed in chapter 9), in a few pages Myron Ebell of the Competitive Enterprise Institute manages to spout at least a dozen misleading claims about global warming.[2] (Remember that before he took on global warming, Ebell made his living denying that secondhand smoke is a health hazard.) As the claims begin to pile up, one's head begins to spin. "Surely he can't be wrong about everything," one thinks in resignation, "at least some of these claims must be right."

Repetition. Keep saying over and over, in every conceivable venue, that smoking is not a health hazard. That was the strategy of Big Tobacco, until the medical evidence grew so strong that to continue saying it risked a lawsuit. The public will believe that something said so often must be true—otherwise how could "they" keep saying it? So keep repeating, over and over, that "in the 1970s scientists forecast a coming ice age," or that "climate models don't work."

Return to the Beginning. Like an army in skillful retreat, the deniers throw up a succession of claims and fall back from one line of defense to the next as scientists refute each in turn. Then they start over:

"The earth is not warming."

"All right, it is warming, but the Sun is the cause."

"Well then, humans are the cause, but it doesn't matter, because warming will do no harm. More carbon dioxide will actually be beneficial. More crops will grow."

"Admittedly, global warming could turn out to be harmful, but we can do nothing about it."

"Sure, we could do something about global warming, but the cost would be too great. We have more pressing problems here and now, like AIDS and poverty."

"We might be able to afford to address global warming some-day, but we need to wait for sound science, new technologies, and geoengineering."

"The earth is not warming. Global warming ended in 1998; it was never a crisis."

Manufacture Doubt. In a criminal trial, the prosecutor has to convince a jury that the accused is guilty "beyond a reasonable doubt." In a civil trial, the standard is "preponderance of the evidence." Global warming deniers demand that before we act to prevent or mitigate global warm-ing, the evidence meet the first standard—be equivalent to proof. But precaution in the face of so great a threat ought to lead us to act when the evidence meets the second standard.

In either case, a defense attorney has to prove nothing—only sow enough doubt to weaken the prosecution's case. Deniers have a huge advantage over scientists, since all they need do is confuse the public enough to make delay seem reasonable. A clever defense attorney can say anything that the judge will allow, whether true or not. Inherently cautious and inclined to wait for stronger evidence, scientists are like miscast prosecutors whose mores and natural reticence play into denier hands.

Ad Hominem. The "To the Man" argument attacks not ideas, but character and reliability. This is why prosecutors try to get in front of the jury any previous criminal convictions that a defendant may have. It is why, when Harrison Schmitt referred to the findings of the IPCC that global warming is real and dangerous, he added with a mocking sneer, "Surprise, surprise." It is why Monckton called those who accept global warming "moaning minnies . . . greens too yellow to admit they are red." It is why the deniers accuse Al Gore of having political or fi-nancial motives. It is why they attack not only James Hansen's science, but his motives. Discredit the man or woman, and you have discredited their argument.

Avoid Accountability: Scientists are accountable for what they say and write; deniers are not. Scientists who make claims not supported by the evidence will not get their articles past peer-review. A scientist who makes a dubious claim in a conference presentation will immediately find hands in the air and pointed questions challenging the assertion.

Speakers at a Heartland Institute conference repeat claims long proven to be false and receive nothing but approbation from their audience. In speeches before Congress, Sen. James Inhofe and Representative Paul Broun say on the record that global warming is a hoax and not only never have to back up their statement with evidence, but receive the applause of their colleagues.

Absence of evidence is evidence of conspiracy. Conspiracy theorists of all stripes regard a lack of evidence of conspiracy as proof of a successful cover-up. The lack of credible evidence that Israel or the U.S. government was behind a 9/11 conspiracy, or that the KGB, or Castro, or whomever, killed John F. Kennedy, merely shows how duplicitous each is, take your pick. That the stolen e-mails do not inculpate James Hansen serves only to show how clever the NASA scientist was at covering his tracks.

Overwhelming evidence is evidence of conspiracy. Since deniers are right and scientists are wrong, that the scientific facts *appear* to offer overwhelming support for global warming merely shows how thorough and successful the conspiracy has been.

Code Words. In a memorandum prepared for Republican clients, consultant Frank Luntz wrote that, "Facts only become relevant when the public is receptive and willing to listen."[3] One of the first actions of the administration of George W. Bush was to rescind Bill Clinton's decision to lower the standard for arsenic in drinking water from 100 to 50 ppm, a decision that proved hard to explain to the public. The problem, Luntz said, was not the fact of the rescission, but in the way the administration had described it. The administration should have said, "Based on sound science, the government's standard is that there should be no more than 50 parts of arsenic per million."

After arsenic, Luntz turned to global warming—or rather, to climate change. In a section in his memo called "Winning the Global Warming Debate," he advises his clients to "Continue to make the lack of scientific certainty a primary issue . . . emphasize the importance of '*acting* only with all the facts.' . . . The most important principle is your commitment to sound science. . . . The scientific debate is closing [against us] but . . . there is still a window of opportunity to challenge the science."

As the first of his nine principles of environmental policy and global warming, Luntz wrote, "Sound science must be our guide in choosing which problems to tackle and how to approach them." President George W. Bush used "sound science" repeatedly, almost always to describe a future state of knowledge for which he was waiting before acting. Calls for sound science pervade denier statements and speeches. Showing how far Luntz's advice has reached, environmentalists and liberals also use the expression. They shouldn't. For sound science to exist, there must be unsound science, or "junk science," as some of the loudest deniers call it. But that is a false and manipulative dichotomy. Base your conclusions about science on the peer-reviewed literature. Then there is no such thing as unsound science and no such thing as sound science. There is just science.

Luntz advised his audience to use the phrase "'Climate change,'" which he said "is less frightening than 'global warming.'" As one focus group participant noted, "climate change 'sounds like you're going from Pittsburgh to Fort Lauderdale.'" While global warming has catastrophic connotations attached to it, "climate change suggests a more controllable and less emotional challenge."

In a 2007 interview on NPR's *Fresh Air* with Terry Gross, Luntz redefined "Orwellian" as a compliment, saying that, "To be 'Orwellian' is to speak with absolute clarity, to be succinct, to explain what the event is, to talk about what triggers something happening . . . and to do so without any pejorative whatsoever."[4] Orwell, who invented Newspeak, where words are "deliberately constructed for political purposes: words, that is to say, which not only had in every case a political implication, but were intended to impose a desirable mental attitude upon the person using them," would have seen through Luntz in an instant.[5]

The Big Lie. In its 2009 Declaration, the Heartland Institute said bluntly that "there is no convincing evidence that CO_2 emissions from modern industrial activity has in the past, is now, or will in the future cause catastrophic climate change." But scientists do not claim that global warming has in the past or now caused "catastrophic climate change"; it is the future that worries them.

In the appendix to *State of Fear*, Michael Crichton was more subtle. While denying that he did so, Crichton likened global warming to Nazi eugenics and Soviet Lysenkoism, as though the logic of the comparison is self-evident. But to equate today's climate scientists with the Lysenkoists is the opposite of the truth.

In the 1920s, to increase crop production, Soviet leaders forced farmers to give up their land to large collective farms. The farmers grew restive, production fell, and in the "breadbasket of Europe," millions starved. Then came the Rasputin of Soviet science, Trofim Denisovitch

Lysenko, who claimed he could make wheat flower earlier, putting more farmers to work and increasing grain production. That was biologically possible, but Lysenko went further to claim that the *offspring* of the "vernalized" wheat would also flower earlier, as though a parent who lifts weights will have more muscular children. Genetics showed instead that characteristics are passed by genes, which are unaffected by traits the parent has acquired. Lysenko denounced geneticists as bourgeois, fascist, pseudoscientists: "fly-lovers and people haters."[6]

Lysenko's image as the peasant genius outwitting the world's biologists dovetailed perfectly with Soviet mythology. In 1938 the authorities placed him in charge of the Academy of Agricultural Sciences, and in 1948 they fired all geneticists and outlawed dissent from Lysenkoism. Purges sent his opponents to prison, some to the executioner. Lysenko was personally responsible for the imprisonment and death by malnutrition of the great Soviet biologist Nikolai Vavilov. Lysenkoism was not a Stalinist aberration; it ruled Soviet biology until the ouster of Khrushchev in the 1960s.

The parallels between the Lysenkoists and the global warming deniers are many. The deniers treat the scientists of the Intergovernmental Panel on Climate Change with contempt, as though it were common knowledge that they are corrupt. Scientist-deniers vilify mainstream scientists like James Hansen; others demand that NASA fire him. After Hansen condemned a presentation that Monckton was to make to the Kentucky state legislature, Monckton wrote the head of NASA accusing Hansen of having financial ties to Al Gore and demanding an investigation.[7] Monckton calls climate scientists evil and likens them to war criminal Radovan Karadžić.

Lysenko accused his scientific opponents of trying to "wreck" the Soviet economy. Today's deniers accuse climate scientists of wanting to transfer money and power from the people to the government, thus helping to bring down "industrialization and development and capitalism and the Western way."

Instead of conducting experiments that would prove his theories, Lysenko used questionnaires from farmers fearful of a one-way ticket to the Gulag. Instead of doing research, the global warming deniers use petitions.

The Soviet media endorsed Lysenko and condemned his opponents, *Pravda* saying that he had "solved the problem of fertilizing the fields without fertilizer and minerals."[8] Today, right-wing American media like Fox News and the *Wall Street Journal* ridicule scientists and provide the deniers with a platform to say whatever they like without fear of contradiction.

In Lysenkoism, the Soviet State denounced biological science and made the denial of genetics state policy; today's deniers urge our government to reject climate science and make the denial of global warming state policy.

The deniers use many other deceitful tactics, but this list is illustrative. Sadly, the modern media, even the most prestigious national newspapers, aid the deniers. Instead of providing the information that readers and viewers need to decide whom to trust, the media "teach the controversy." If the media are going to quote an Ebell or a Singer, the reporter needs to make it clear that those two have a long history of denial on a variety of topics—that Ebell, for example, has no scientific credentials of any kind, that both have received money from industry to support their denial, and that both belong to organizations with an avowed mission of showing that global warming is false. After all, reporters have to disclose their own conflicts. Should they not reveal the obvious conflicts of those they interview?

To Roll Back Industrial Society

This brings us to the question of *why* the deniers repudiate the overwhelming scientific consensus on global warming. What motivates them?

A few are simply contrarian by nature, delighting in pricking the majority and reveling in their reputation as free-thinkers unafraid to be different. Every field of science has had contrarians who threw out provocative ideas, most of which, but not all, turned out to be wrong. But even wrong ideas can advance science, stimulating others to conduct new experiments and ask new questions that lead to important discoveries. Some obstinate scientists play a valuable role as Devil's Advocate, but others have simply been wrong—repeatedly. Freeman Dyson has advocated carbon-eating trees, nuclear-powered space rockets, and even entertained the idea that the United States might win the Vietnam War by dropping tactical nuclear weapons. If Dyson were to accept global warming, no one would pay any attention to him, since he would then merely be one of tens of thousands contributing to the scientific consensus, almost any one of whom would know as much as he. By denying global warming, Dyson lands on the cover of the *New York Times Magazine* and wins an award for intellectual courage.

Then there are the professional scientist-deniers. Find a topic that an industry opposes and for a fee these apostates will write books and articles, appear on talk shows, testify before Congress, and spout the industry line. Collectively, those profiled in this book have denied science in order to oppose government regulation of acid rain, CFCs, environmental mercury, fast foods, fossil fuel combustion, pesticides, second-hand smoke, and more. Ask where they stand on one of these regulatory issues and you will know there they stand on them all. The deniers proclaim their support for individual rights and "Liberty," then foster the agendas not of individuals, but of polluting corporations. Here's one example. Between 1998 and 2005, the Annapolis Center for Science-Based Public Policy received more than three-quarters of a million dollars from ExxonMobil. Harrison Schmitt, who labels global warming a threat to liberty, served as the center's president from 1994 to 1998. The Annapolis Center used the funds in part to question the connection between air pollution and asthma, the health effects of environmental mercury, and the dangers of pesticides on food.

The trait that most deniers seem to have in common is that they are economic and political libertarians, opposed to government regulation of every kind. The *Oxford English Dictionary* defines *libertarian* as "A person who believes the role of the government should be limited to upholding individual rights, and who therefore opposes government regulation of economic or social affairs." Richard Lindzen, to take one example, "sees climate advocates as wanting 'to roll back industrial society' and an excuse 'to redistribute global wealth.'"[9]

To summarize their traits, deniers:

- Dispute a scientific consensus that is at least twenty years old
- Cite an online petition as proof of the absence of consensus
- Manufacture doubt by cherry-picking and other methods
- With the exception of one accomplished scientist, almost never do research or publish in peer-reviewed journals
- Derogate peer-review as designed to quash new ideas
- Portray themselves as beleaguered, gutsy truth-tellers denied the research grants they need to show that mainstream scientists are wrong

- Conduct dissident conferences with all the trappings of legitimate scientific meetings
- Accuse scientists and the government of a global conspiracy
- Receive support from the media and heads of state
- Call scientists Nazis and murderers

Though this list fits the global warming deniers perfectly, in fact I based it on the AIDS denialists described by Seth Kalichman in his important book, *Denying AIDS: Conspiracy Theories, Pseudoscience, and Human Tragedy*. One could come up with a similar list for evolution deniers, vaccine deniers (those who claim that vaccines cause autism), and others of their ilk. Indeed, as Debora MacKenzie observes in *New Scientist*, all denier movements "set themselves up as courageous underdogs fighting a corrupt elite engaged in a conspiracy to suppress the truth or foist a malicious lie on ordinary people. This conspiracy is usually claimed to be promoting a sinister agenda: the nanny state, takeover of the world economy, government power over individuals, financial gain, atheism."[10]

We cannot really know the minds of individual deniers. We cannot know even our own minds, much less the minds of others, especially those who seem so impervious to reason. What we do see clearly is that science denial begins and ends not with science but with ideology, not with facts but with belief systems.

Escalating Tactics

In the last few years, the tactics of global warming deniers have evolved to mimic those of the creationists. Even though they have almost always lost in court, creationists have been able to gain control not only of local school boards, but of state boards of education and even state legislatures. One of their most successful strategies has been to appeal to the good old American notion of fair play. When an issue is in dispute, should we not hear from both sides? This not only seems to represent common sense, it is the foundation of parliamentary democracy and our court system. But in science, almost never do two sides deserve equal attention. We do not give equal time to HIV denial, vaccine denial, moon-landing denial, and round-earth denial. Nor to Ptolemy's astronomy, though between 15 and 20 percent of those polled in Western nations believe that the Sun travels around the earth.[1]

Creationists argue that instead of teaching only biology in the biology classroom, schools should introduce "creation science," so that students can learn from the "debate." They have resurrected this strategy with their latest modification, intelligent design. According to its primary champion, the Discovery Institute, "The theory of intelligent design holds that certain features of the universe and of living things are best explained by an intelligent cause, not an undirected process such as natural selection."[2] Intelligent design repeats the argument William Paley made in 1802: if you find a watch lying in your path, you know that a designer—the watchmaker—built it. Since we find even more complex objects in Nature—the eye, for instance—someone must have made them too and that someone can only have been God.

In a 2005 ruling, Pennsylvania Judge John E. Jones described the "breathtaking inanity" of school board members who wanted to promote intelligent design, accusing several of lying to conceal their true motive: to promote religion.[3]

State legislatures have now begun to follow the creationist model by demanding equal time for global warming denial. In 2008, Louisiana's Governor Bobby Jindal signed a law "to allow and assist teachers and school administrators to create and foster an environment that promotes critical thinking skills, logical analysis, and open and objective discussion of scientific theories being studied including, but not limited to, evolution, the origins of life, global warming, and human cloning."[4] The Oklahoma legislature took up a similar bill in 2009, but failed to pass it. Bills linking evolution and global warming were proposed in the legislatures of Kentucky and South Carolina, but died in committee.[5]

The next logical step is for state legislatures to single out global warming, without bothering to lump it together with evolution. In 2010 the South Dakota legislature passed House Concurrent Resolution No. 1009, which begins with a series of "whereases" (quoted exactly below) taken straight from the denier playbook:[6]

The earth has been cooling for the last eight years despite small increases in anthropogenic carbon dioxide;

There is no evidence of atmospheric warming in the troposphere where the majority of warming would be taking place;

During the Little Climatic Optimum, Erik the Red settled Greenland where they farmed and raised dairy cattle. Today, ninety percent of Greenland is covered by massive ice sheets, in many places more than two miles thick;

The polar ice cap is subject to shifting warm water currents and the break-up of ice by high wind events. Many oceanographers believe this to be the major cause of melting polar ice, not atmospheric warming;

Carbon dioxide is not a pollutant but rather a highly beneficial ingredient for all plant life on earth. Many scientists refer to carbon dioxide as "the gas of life";

More than 31,000 American scientists collectively signed a petition to President Obama stating: "There is no convincing scientific evidence that human release of carbon dioxide, or methane, or other greenhouse gasses is causing or will, in the foreseeable future, cause catastrophic heating of the earth's atmosphere and disruption of the earth's climate. Moreover, there is substantial scientific evidence that increases in atmospheric carbon dioxide will produce many beneficial effects on the natural plant and animal environments of the earth.

Therefore, the bill urges, instruction in the public schools relating to global warming should include the following:

1. That global warming is a scientific theory rather than a proven fact
2. That there are a variety of climatological, meteorological, astrological, thermological, cosmological, and ecological dynamics that can effect [sic] world weather phenomena and that the significance and interrelativity of these factors is largely speculative; and
3. That the debate on global warming has subsumed political and philosophical viewpoints which have complicated and prejudiced the scientific investigation of global warming phenomena

As though, like pornography, global warming might taint young minds, the South Dakota legislature urges that "all instruction on the theory of global warming be appropriate to the age and academic development of the student and to the prevailing classroom circumstances."

Monkey Trials

The next strategy of the deniers is to use the courts. A seminal moment in the history of creationism was the 1925 Tennessee "Monkey Trial" of teacher John Scopes for violating a state law forbidding teachers "to teach any theory that denies the story of the Divine Creation of man as taught in the Bible." Scopes was convicted, but the state supreme court overturned on a technicality. By most estimates, the national exposure of the Scopes trial weakened the creationists' case.

In August 2009, as the Environmental Protection Agency got ready to rule that CO_2 is a public danger, the U.S. Chamber of Commerce urged the agency to hold a "trial-style public hearing" on global warming so as "to make a fully informed, transparent decision with scientific integrity based on the actual record of the science." The hearing would be "the Scopes monkey trial of the 21st century," chortled William Kovacs, the chamber's senior vice president for environment, technology, and regulatory affairs. Kovacs soon withdrew his remarks, writing, "My 'Scopes monkey' analogy was inappropriate and detracted from my ability to effectively convey the Chamber's position on this important issue," words that have the cast of a lawyer's pen.

Global warming deniers have already had a junior version of the Monkey Trial. In 2007 a British lorry driver and parent of children at state school sued the UK's Department for Education and Skills because it had distributed Al Gore's film, *An Inconvenient Truth*, to every secondary school along with a guidance note for teachers. The suit claimed that the film violated a provision in British law against "political indoctrination" and also failed to live up to the "duty to secure balanced treatment of political issues." The deniers needed to convince not a jury, but just one judge. But like the Scopes trial, the strategy backfired, Mr. Justice Burton finding that the film "is substantially founded upon scientific research and fact," advancing "four main scientific hypotheses, each of which is very well supported by research published in respected, peer-reviewed journals and accords with the latest conclusions of the Intergovernmental Panel on Climate Change." To paraphrase these four: temperatures are rising and likely to continue to rise; human emissions of CO_2 are the cause; climate change is dangerous; individuals and governments can reduce or mitigate its effect.

But the courtroom may prove an uncomfortable place for global warming deniers, as those who have suffered from the effects of rising seas and other climate-induced calamities sue Big Coal and Big Oil, partly on the grounds that companies like ExxonMobil claimed that global warming was false even as their own internal documents showed it to be true. Big Tobacco lost hundreds of billons of dollars not because it made a lethal product, but because it lied.

Old Virginia Home

As we saw with Senator Inhofe's attempt to make criminals out of seventeen climate scientists, elected officials have the power to harass and condemn scientists without benefit of trial. Consider Ken Cuccinelli II, attorney general of Virginia. In his first 100 days in office, Cuccinelli wrote university presidents advising them to rescind policies protecting gays and lesbians from discrimination, covered with armor plate the bare breasts of Roman goddess Virtus on the Virginia state seal, and filed a petition with the Environmental Protection Agency asking it to reconsider its finding that CO_2 endangers health.

On April 23, 2010, Cuccinelli demanded that the University of Virginia divulge a broad range of documents to determine whether Michael Mann had defrauded Virginia taxpayers in seeking research grants while employed by the university in 1999. The grounds for the April petition were the e-mails stolen from the Climate Research Unit of the University of East Anglia, which Cuccinelli said showed "scientists using faulty data to support the notion of manmade global warming." The erroneous data had influenced the IPCC, he claimed, rendering its findings "unreliable, unverifiable and doctored." Remember that by April 23, 2010, three independent panels had exonerated Mann and the University of East Anglia. (Later, three more panels did the same.)

Cuccinelli's demand makes chilling reading. It "commands" the Rector of the university to produce all documents and other materials "from January 1, 1999, through the present date," as well as "Any documents prepared during this time period, or before this time period but which relate thereto, . . ." An attachment spells out just what "any documents" means:[7]

> The scope of this Civil Investigative Demand is to reach any and all data, documents and things in your possession, including but not limited to, those stored on any computer, hard drive, desktop, laptop, file server, database server, e-mail servers or other systems where data was transmitted or stored on purpose, or as a result of transient use of a system or application in the course of day-to-day research or product processing that is owned or contracted for by you [the University] or any of your officers, managers, employees, agents, board members, academic departments, divisions, programs, IT department, contractors or other representatives.

There follows a list of thirty-three individuals, most of them PhD climate scientists, including some critics of Mann's work, as well as "all research assistants, secretaries or administrative staff with whom Dr. Mann worked while he was at the University of Virginia." For the persons named, the University is to produce

> All documents that constitute or are in any way related to correspondence, messages, or e-mails sent to or received by Dr. Michael Mann

All documents that constitute or that are in any way related to correspondence, messages, or e-mails sent from Dr. Michael Mann

All documents that constitute or that are in any way related to correspondence, messages, or e-mails sent to or from Dr. Michael Mann that reference [the list or names follows]

The most ominous section of the demand describes what the University is supposed to do if "any document requested was, but is no longer in your possession, subject to your control, or in existence." In that case, the University is to describe for each such document:

(a) the type of document
(b) whether it is missing, lost, has been destroyed, or has been transferred to the possession, custody, or control of other persons
(c) the circumstances surrounding, and the authorization for, the disposition
(d) the date or approximate date of the disposition
(e) the identity of all as having knowledge of the circumstances described in (c) above; and
(f) the identity of all persons having knowledge of the document's contents

Cuccinelli's demand includes words that might have surprised Kafka himself:

All uses of the conjunctive should be interpreted as including the disjunctive and vice versa in order to bring within the scope of this CID any information or documents that might otherwise be construed to be outside of the scope.

To its credit, The University of Virginia's Faculty Executive Council responded:

[Cuccinelli's] action and the potential threat of legal prosecution of scientific endeavor that has satisfied peer-review standards send a chilling message to scientists engaged in basic research involving Earth's climate and indeed to scholars in any discipline. Such actions directly threaten academic freedom and, thus, our ability to generate the knowledge upon which informed public policy relies.[8]

On May 27, 2010, the university filed a court challenge to Cuccinelli's demand, which President John T. Casteen III said "has sent a chill through the Commonwealth's colleges and universities—a chill that has reached across the country and attracted the attention of all of higher education."[9]

On August 30, 2010, the Sixteenth Judicial Court of the Commonwealth of Virginia issued its decision on the university's appeal.[10] It noted first that Cuccinelli argues that the Virginia Attorney General "has unbridled discretion to say he believes the University does have relevant material and that the Court does not have the ability to review any requirement he has 'reason to believe' or not." But, "The Court disagrees. In order for the Attorney General to have 'reason to believe,' he has to have some objective basis to issue a civil investigative demand." For "objective basis," read "evidence." Evidently, Cuccinelli had none.

Second, "What the Attorney General suspects that Dr. Mann did that was false or fraudulent . . . is simply not stated." The Virginia Deputy Attorney General provided a brief, but Judge Paul B. Peatross Jr. found it wanting, saying that although he "understands the controversy regarding Dr. Mann's work . . . it is not clear what he did that was misleading, false, or fraudulent in obtaining funds from the Commonwealth of Virginia." Finally, the Judge noted that four of the five research grants cited in the demand were federal, not state grants, and thus were outside its jurisdiction. As to the fifth, the University of Virginia made the grant in 2001, two years before the Virginia Fraud Against Taxpayers Act became law.

The court "set aside" the CID without prejudice, meaning that Cuccinelli could try again. In October 2010, responding to the judge's objections, Cuccinelli filed a new CID.[11] Now he gave as the reasons for the demand that Mann had published two papers "which have come under significant criticism" and which "he knew or should have known contained false information, unsubstantiated claims, or were otherwise misleading." Some of the conclusions, the demand continued, "demonstrate a complete lack of rigor meaning that the result reported lacked statistical significance without a specific statement to that effect." An appendix cites extensively from the Climategate e-mails. In a separate action, in December 2010 the Attorney General appealed Judge Peatross's original denial to the Virginia Supreme Court, thus putting his efforts on two parallel tracks.

When senators and attorneys general want to criminalize scientists, when broadcasters call for a public flogging of scientists or urge them to commit suicide, should we laugh them off? Before we do, we might consider some recent history.

Recall American terrorist Timothy McVeigh, who in April 1995 blew up the Murrah federal building in Oklahoma City, killing 168 people and injuring 450. McVeigh was said to be seeking revenge for the FBI siege at Waco, Texas, two years earlier. Or more recently, think back to February 2010, when a man angry at the U.S. Internal Revenue Service crashed his plane into a Texas office building where nearly 200 IRS employees worked, starting a conflagration that sent smoke high above the seven-story building. Suppose that instead of a federal building or IRS office, the target had been EPA workers or a university climate science laboratory. Why not? Such attacks happened in the 1960s. Read the global warming denier blogs, and it is hard to imagine that anyone is angrier at the IRS than global warming deniers are angry at climate scientists and the EPA.

If deniers can vilify individual scientists and neutralize the field of climate science simply because of ideology and a conspiracy theory, what will be the next field of science—or art, or history, or literature, or medicine—that some group chooses to denounce?

Earning Trust

It comes down to trust. Global warming deniers ask us to trust them and to distrust scientists individually and collectively. But the American public has always trusted scientists, and for good reason. Should it stop doing so now, when we need science more than ever?

A poll by the Pew Research Center in July 2009 found that 84 percent of respondents believed that science has had a positive effect on society. On the question of which professions contribute "a lot" to society's well-being, the poll ranked the military first at 84 percent, followed by teachers at 77 percent, scientists at 70 percent, and medical doctors at 69 percent. Clergy came in at only 40 percent. Holding up the bottom of the list were lawyers at 23 percent and business executives at 21 percent.

As reported in *Science and Engineering Indicators 2010*, published by the National Science Foundation, the public has more confidence in the science community than in either the military or medical communities. On the question of who should have the most influence in decisions about global warming, environmental scientists came in at 85 percent, compared with 50 percent for elected officials and only 32 percent for business leaders.

Why does the public trust scientists? Any mature person knows the many scientific and medical advances that have occurred since their youth. We know that without the progress brought by science and medicine over the last two centuries, our lives would be hard, inconvenient, and short.

In high school we learn that one reason to trust scientists is because their work is self-correcting. Peer review means that one can trust scientific articles far more than the statements of politicians, journalists, and bloggers. A few erroneous claims do make it through peer review; but then other scientists expose the errors when they cannot replicate the results. Remember the case of cold fusion, which two scientists announced at a press conference in March 1989 and which, only two months later, other scientists thoroughly debunked.

Many Americans have known a scientist, perhaps a high school teacher or a neighbor down the street. Scientists tend to be well educated, serious, and hard-working, sometimes to the point of being workaholics. They may be on the introverted side; they show no evidence of being more interested in politics or ideology than the average American. Does it make sense to believe that tens of thousands of scientists would be so deeply and secretly committed to bringing down capitalism and the American way of life that they would spend years beyond their undergraduate degrees working to receive master's and PhD degrees, then go to work in a government laboratory or university, plying the deep oceans, forbidding deserts, icy poles, and torrid jungles, all for far less money than they could have made in industry, all the while biding their time like a Russian sleeper agent in an old spy novel?

Scientists tend to be independent and to resist authority. That is why you are apt to find them in the laboratory or field, as far as possible from the prying eyes of a supervisor. Indeed, many do not recognize that they have a supervisor. Anyone who believes that he could organize thousands of scientists into a conspiracy has never attended a single faculty meeting.

By trusting science, those of us in the Western world have achieved a standard of living and a life expectancy unimaginable even two centuries ago. There is not a shred of evidence that scientists have lied about global warming. The positive evidence for global warming is overwhelming and irrefutable. To distrust science and scientists now, for no good reason and at the very moment they warn of the greatest threat ever to face humanity, is to abandon reason and our uniquely human ability to imagine the future.

The clock is ticking. Our leaders do not have the luxury of waiting decades to find out if scientists are right about global warming. By the time there is sufficient certainty to satisfy the deniers, it will be far too late to limit the worst effects of global warming. We do not really even know how bad those worst effects might be.

We could have acted at the turn of the century to begin to reduce carbon emissions: the science was clear enough. Now, a decade later, we still have not acted. In another two decades, humanity will have passed the point at which CO_2 can be limited to 450 ppm, the highest level that may—we cannot be sure—avoid runaway global warming.

It is clear that our political leaders are unwilling to get ahead of public opinion. But public opinion can change if people will think for themselves and look at the evidence for global warming and at the deceitful and mendacious claims of the industry of denial. When politicians are unable to win election without accepting science and taking a stand to limit carbon emissions, those emissions will begin to decline.

We can fault our leaders, but whether we trust science and act in time to avoid the worst dangers of global warming is really not up to them. It is up to us. To you.

Appendix

The science academies of the following countries or regions have issued statements accepting global warming.[1]

Australia	Indonesia	Senegal
Belgium	Ireland	South Africa
Brazil	Italy	Sudan
Cameroon	Japan	Sweden
Canada	Kenya	Switzerland
Caribbean	Madagasgar	Tanzania
China	Malaysia	Uganda
France	Mexico	United Kingdom
Germany	Nigeria	United States
Ghana	New Zealand	Zambia
India	Russia	Zimbabwe

National and international organizations from nearly every field of science have issued statements accepting global warming. No scientific organization, which excludes denier websites and front groups, has issued a statement rejecting the basic findings of human-caused global warming.

American Academy of Pediatrics	American Association for the Advancement of Science
American Association of Petroleum Geologists	American Association of State Climatologists
American Association of Wildlife Veterinarians	American Astronomical Society
American Chemical Society	American College of Preventive Medicine
American Geological Institute	American Geophysical Union
American Institute of Physics	American Institute of Biological Scientists

American Institute of Professional
Geologists

American Medical Association

American Meteorological Society

American Physical Society

American Public Health
Association

American Quaternary Association

American Society for
Microbiology

American Statistical Association

Arctic Climate Impact
Assessment

Australian Coral Reef Society

Australian Institute of Physics

Australian Medical Association

Australian Meteorological and
Oceanographic Society

Canadian Federation of Earth Sciences

Canadian Foundation for Climate
and Atmospheric Sciences

Canadian Meteorological and
Oceanographic Society

Engineers Australia

European Academy of Sciences and Arts

European Federation of
Geologists

European Geosciences Union

European Physical Society

European Science Foundation

Federation of Australian Scientific
and Technological Societies

Geological Society of America

Geological Society of Australia

Geological Society of London

Institute of Biology (UK)

Institute of Professional Engineers (NZ)

InterAcademy Council

Intergovernmental Panel on Climate Change

International Association for
Great Lakes Research

International Council of Academies of
Engineering and Technological Sciences

International Council on Science

International Union for Quaternary Research

International Union of Geodesy
and Geophysics

NASA's Goddard Institute of Space Studies

National Association of
Geoscience Teachers

National Center for Atmospheric Research

National Oceanic and
Atmospheric Administration

National Research Council

Network of African Science
Academies

Pew Center on Climate Change

Polish Academy of Sciences

Royal Meteorological Society (UK)

Royal Society of New Zealand

Society of American Foresters

State of the Canadian
Cryosphere

The Wildlife Society (international)

Union of Concerned Scientists

US Geological Survey

Woods Hole Oceanographic
Institute

Woods Hole Research Center

World Federation of Public
Health Associations

World Health Organization

World Meteorological
Organization

World Wildlife Fund

Notes

(All websites visited between May 27 and July 1, 2010)

Introduction

1. Gerald Warner, "Government 'Scientific Advisers': Who Needs These Nuts in White Coats?" *The Telegraph* (1/16/2010): *see* http://blogs.telegraph.co.uk/news/geraldwarner/100022443/government-scientific-advisers-who-needs-these-nuts-in-white-coats/.

2. Ben Bergmann, "Rep. Broun Receives Applause on the House Floor for Calling Global Warming a 'Hoax,' " *Climate Progress* (6/26/2009): *see* http://climateprogress.org/2009/06/26/rep-broun-receives-applause-on-the-house-floor-for-calling-global-warming-a-'hoax'/.

3. Jim Inhofe, "Welcome to the Inhofe EPW Press Office," United States Senate Environment and Public Works Committee: *see* www.epw.senate.gov/public/index.cfm?FuseAction=Minority.WelcomeMessage.

4. Peter Gleick, "The Best Argument Against Global Warming," *San Francisco Chronicle* (3/11/2010): *see* www.sfgate.com/cgi-bin/blogs/gleick/detail?blogid=104&entry_id=58962.

5. David Michaels, *Doubt Is Their Product: How Industry's Assault on Science Threatens Your Health*.

6. Seth C. Kalichman, *Denying Aids: Conspiracy Theories, Pseudoscience, and Human Tragedy*.

7. Oxford University Press, *The Oxford Dictionary of Quotations*, 3d ed., 24.

8. Thomas Henry Huxley, *Selected Essays and Addresses of Thomas Henry Huxley*, ed. Philo Melvyn Buck, Jr., 49.

9. In his important book, *Denying Aids*, Seth Kalichman makes a good case for his decision to call those who deny that HIV causes AIDS, "denialists." One reason that he does so is that in medicine "denial" can refer to the first of Elizabeth Kubler-Ross's

five stages of coping strategies. Another is that Kalichman believes that *denialist* "better represents the psychological process of malignant denial that is inherent in some denialism."

1. Science and Potemkin Science

1. American Geophysical Union, "Mission, Vision, and Goals" (6/07/2010): *see* www.agu.org/about/mission.shtml.
2. Philip Shabecoff, "Global Warming Has Begun, Expert Tells Senate," *New York Times* (6/24/2008).
3. *See* www.earthsky.org/.
4. Wallace S. Broecker, "Climatic Change: Are We on the Brink of a Pronounced Global Warming?" *Science* 189 (1975): 460–63
5. Wallace S. Broecker and Robert Kunzig, *Fixing Climate: What Past Climate Changes Reveal About the Current Threat—and How to Counter It.*
6. Wallace S. Broecker, "Shifting Rainfall: A Paleo Perspective," *EOS Trans. AGU* 89(53) (Fall Mtg. Supp., Abstract UA22A-05, 2008).
7. *See* www.heartland.org/events/NewYork09/index.html.
8. Presentation at the 2009 Heartland Institute Conference by former Senator Harrison Schmitt.
9. *See* www.heartland.org/about/.
10. *See* www.heartland.org/events/NewYork09/proceedings.html.
11. Lindzen, Richard, "Climate Alarm: What We Are Up Against, and What To Do," *The Heartland Institute* (3/2009): *see* www.heartland.org/full/24841/Climate _Alarm_What_We_Are_Up_Against_and_What_to_Do.html.
12. *See* www.heartland.org/events/NewYork09/proceedings.html.
13. Both are available at www.sepp.org/.
14. *See* www.heartland.org/policybot/.

2. Adventures in Denierland

1. *See* www.heartland.org/events/NewYork09/proceedings.html.
2. *See* www.climatescienceinternational.org/.
3. *See* www.climatescienceinternational.org/.
4. *See* http://nzclimatescience.net/.
5. National Research Council, *Climate Change Science: An Analysis of Some Key Questions.*
6. Andrew C. Revkin, "Climate Expert Says NASA Tried to Silence Him," *New York Times* (1/27/2006).
7. *See* www.climatesciencewatch.org/.
8. M. E. Mann, R. S. Bradley, and M. K. Hughes. "Global-Scale Temperature Patterns and Climate Forcing over the Past Six Centuries." *Nature* 392 (1998): 779–87.

9. *See* www.fraserinstitute.org/researchandpublications/publications/3184.aspx.

10. "Fraser Institute fires off a damp squib," *RealClimate* (2/03/2007): *see* www
.realclimate.org/index.php/archives/2007/02/fraser-institute-fires-off-a-damp-
squib/.

11. Oregon Institute of Science and Medicine: *see* www.oism.org/.

12. "Petition Project," Oregon Institute of Science and Medicine: *see* www.oism
.org/pproject/.

13. "The great Petition Fraud," Greenfyre: *see* http://greenfyre.wordpress.com/
denier-vs-skeptic/denier-myths-debunked/climate-denial-crock-of-the-week/#great.

14. H. J. Hebert, "Jokers Add Fake Names to Warming Petition," *Seattle Times*
(5/1/1998)

3. The Evidence for Consensus

1. R. Matthews, "Leading Scientific Journals 'Are Censoring Debate on Global
Warming.'" *The Telegraph* (5/1/2005).

2. *See* http://scienceblogs.com/deltoid/2006/10/peiser_admits_he_was_97%wrong
.php.

3. Christopher Monckton, " 'Consensus'? What 'Consensus'? Among Climate
Scientists, the Debate Is Not Over," *see* http://scienceandpublicpolicy.org/monckton/
consensuswhatconsensusamongclimatescientiststhedebateisnotover.html.

4. "What does Naomi Oreskes' study on consensus show?" *Skeptical Science*: *see*
www.skepticalscience.com/naomi-oreskes-consensus-on-global-warming.htm.

5. Tim Lambert, "Schulte replies to Oreskes," *Deltoid* (9/9/2007): *see* http://
scienceblogs.com/deltoid/2007/09/schulte_replies_to_oreskes.php.

6. Naomi Oreskes, "The Scientific Consensus on Climate Change," *Science* 306
(2004): 1686.

7. Donald Kennedy, "Editorial—an Unfortunate U-Turn on Carbon," *Science* 291
(2001): 2.

8. Baker, quoted in Maxwell T. Boykoff and Jules M. Boykoff. "Balance as Bias:
Global Warming and the US Prestige Press." *Global Environmental Change: Human
and Policy Dimensions* 14 (2004): 125–36.

9. National Research Council, *Climate Change Science: An Analysis of Some Key
Questions.*

10. The American Association of Petroleum Geologists, Division of Professional
Affairs, Governmental Affairs Committee: *see* http://dpa.aapg.org/gac/index.cfm.

11. Nobel Laureates, "The St. James's Palace Memorandum: 'Action for a Low
Carbon and Equitable Future,'" London, UK, May 26 – 28, 2009.

12. R. Showstack, "Science Academies' Statement on Climate Change." *EOS* 90
(2009): 216. The nations are Brazil, Canada, China, France, Germany, India, Italy,
Japan, Mexico, Russia, South Africa, United Kingdom, and the United States.

13. National Research Council, *Advancing the Science of Climate Change.*

14. "Policy on AGU's Role in Advocacy on Public Issues," American Geophysical
Union (5/1982): *see* www.agu.org/about/governance/policies/advocacy.shtml.

15. "AGU Position Statements," American Geophysical Union: *see* www.agu.org/outreach/science_policy/positions/climate_change2008.shtml.

16. Fairness & Accuracy in Reporting: *see* www.fair.org/.

17. George Will, "Al Gore's Green Guilt," *Washington Post* (9/3/1992).

18. *See* www.fair.org.

19. Peter Doran and Maggie Kendall Zimmerman, "Examining the Scientific Consensus on Climate Change," *EOS* 90 (2009): 22–23.

20. Peter Gleick, and 254 others, "Climate Change and the Integrity of Science," *Science* 328 (2010): 689–90.

21. William R. L. Anderegg, James W. Prall, Jacob Harold, and Stephen H. Schneider, "Expert Credibility in Climate Change," *Proceedings of the National Academy of Sciences of the United States of America* 107.27 (2010): 12107–12109.

22. John Tierney, "The Big City: It's a Sin." *New York Times* (4/9/1995).

23. John Tierney, "Flawed Science Advice for Obama?" *New York Times* (12/19/2008).

24. John Tierney, "Final Report from the Lab," *New York Times* (5/19/2010): *see* http://tierneylab.blogs.nytimes.com/.

25. Michael Crichton, "Aliens Cause Global Warming," *Wall Street Journal* (1/17/2003): *see* http://online.wsj.com/article/SB122603134258207975.html.

26. Remarks to the Commonwealth Club by Michael Crichton, San Francisco, September 15, 2003; *see* http://online.wsj.com/article/SB122603134258207975.html.

27. *See* www.cfa.harvard.edu/~scranmer/SP/cricDhton.html.

28. Albert Einstein, "Does the Inertia of a Body Depend Upon Its Energy Content?" *Annalen der Physik* 18 (1905): 639–41.

29. Crichton, "Aliens Cause Global Warming."

30. Naomi Oreskes, "Science and Public Policy: What's Proof Got to Do with It?" *Environmental Science & Policy* 7 (2004): 369–83.

4. Discovery of Global Warming

1. For a complete and eloquent account of the history of global warming science, see Spencer R. Weart, *The Discovery of Global Warming*. Online at www.aip.org/history/climate/index.html.

2. John Tyndall, "On the Absorption and Radiation of Heat by Gases and Vapours, and on the Physical Connexion of Radiation, Absorption, and Conduction," *Philosophical Magazine* 22 (1861): 169–94, 273–85.

3. Bernard Jaffe, *Crucibles: The Story of Chemistry from Ancient Alchemy to Nuclear Fission*.

4. Svante Arrhenius and H. Borns, *Worlds in the Making: The Evolution of the Universe*, 54.

5. Ibid., 54, 63.

6. Weart, *The Discovery of Global Warming*, 6.

7. G. S. Callendar, "The Artificial Production of Carbon Dioxide and Its Influence on Climate," *Quarterly Journal of the Royal Meteorological Association* 64 (1938): 223.

8. Ibid., 236.

9. Spencer R. Weart, "The Carbon Dioxide Greenhouse Effect," the American Institute of Physics: see www.aip.org/history/climate/co2.htm.

10. Roger Revelle and Hans E. Suess, "Carbon Dioxide Exchange between Atmosphere and Ocean and the Question of an Increase of Atmospheric CO_2 During the Past Decades," *Tellus* 9 (1957): 18–27.

5. The Greenhouse Effect: From Curiosity to Threat

1. President's Science Advisory Committee, Environmental Pollution Panel (White House, 1965), *Restoring the Quality of Our Environment*.

2. Naomi Oreskes, "The Long Consensus on Global Warming," *Washington Post* (2/19/2007).

3. "Bush at 60: 'I Really Do Feel Young,'" *People* (7/7/2010): see www.people.com/people/article/0,26334,1210402,00.html.

4. National Research Council, Committee on Atmospheric Sciences, *Weather and Climate Modification Problems and Prospects: Final Report of the Panel on Weather and Climate Modification*, 88.

5. National Research Council, Climate Research Board, *Carbon Dioxide and Climate: A Scientific Assessment*, viii. Subsequent references to this work are cited in the text.

6. World Climate Conference (1979); see www.wmconnolley.org.uk/sci/iceage/wcc-1979.html.

7. Intergovernmental Panel on Climate Change: see www.ipcc.ch/.

8. Oreskes, "The Long Consensus on Global Warming."

9. Susan Solomon, *Climate Change 2007: The Physical Science Basis; Summary for Policymakers, Technical Summary and Frequently Asked Questions*, 17.

10. Ibid.

6. Global Warming: All You Really Need to Know in One Chart

1. "How do we know that recent CO_2 increases are due to human activities?" *RealClimate* (12/22/2004): www.realclimate.org/index.php/archives/2004/12/how-do-we-know-that-recent-cosub2sub-increases-are-due-to-human-activities-updated/.

7. Tobacco Tactics: The Scientist-Deniers

1. For a more detailed account of the link between tobacco denial and global warming denial, see Naomi Oreskes and E. M. Conway. *Merchants of Doubt: How a Handful of Scientists Obscured the Truth on Issues from Tobacco to Global Warming*.

2. Science and Environmental Policy Project: *see* www.sepp.org/.

3. Singer, quoted in Ross Gelbspan, *The Heat Is On: The Climate Crisis, the Cover-up, the Prescription,* 47.

4. "Policy Declarations," Science and Environmental Policy Project: *see* www.sepp .org/policy_declarations.

5. Ibid.

6. SourceWatch: *see* www.sourcewatch.org.

7. Ibid.

8. Union of Concerned Scientists, *Smoke, Mirrors & Hot Air: How Exxonmobil Uses Big Tobacco's Tactics to Manufacture Uncertainty on Climate Science,* 18.

9. Ibid.

10. S. Fred Singer, "Anthology of 1995's Environmental Myths," *Washington Times,* (2/11/1996).

11. *See* www.sourcewatch.org.

12. ExxonSecrets: *see* www.exxonsecrets.org.

13. Kevin Grandia, "S. Fred Singer and the Global Warming Denial Machine," Desmog Blog (3/28/2008): *see* www.desmogblog.com/s-fred-singer-and-the-global-warming-denial-machine.

14. Quoted in Gelbspan, *The Heat Is On,* 36.

15. Ross Gelbspan, "The Heat Is On: The Warming of the World's Climate Sparks a Blaze of Denial." *Harper's* 291 (1995): 9.

16. Patrick Michaels, "Testimony: Climate Crisis: National Security, Economic, and Public Health Threats," Cato Institute (2/12/2009): *see* www.cato.org/testimony/ct-pm-20090212.html.

17. "Cato Climate Change Ad," Cato Institute: *see* www.cato.org/special/climate change/alternate_version.html.

18. RealClimate: *see* www.realclimate.org.

19. Frederick Seitz and Robert Jastrow, "Do People Cause Global Warming?" The Heartland Institute (12/01/2001): *see* www.heartland.org/policybot/results/812/Do_people_cause_global_warming.html.

20. Mark Hertsgaard, "While Washington Slept." *Vanity Fair* (5/2006).

21. Thomas Borelli, "000200 ACTIVITY REPORT" (3/1/1994), Legacy Tobacco Documents Library, University of California, San Francisco, *see* http://legacy.library .ucsf.edu/tid/yzb65e00.

22. David Malakoff, "Climate Change: Advocacy Mailing Draws Fire," *Science* 280 (1998): 195

23. *See* www.oism.org/pproject/.

24. "Statement of the Council of the National Academy of Sciences Regarding Global Change Petition," National Academy of Sciences (5/20/1998): *see* www8 .nationalacademies.org/onpinews/newsitem.aspx?RecordID=s04201998.

25. Alexander Holtzman, "FRED SEITZ" (8/31/1989), Legacy Tobacco Documents Library, University of Calinfornia, San Fransisco: *see* http://legacy.library.ucsf .edu/tid/hwj53e00.

26. "The Leipzig Declaration on Global Climate Change," Science & Environmental Policy Project: *see* www.sepp.org/policy%20declarations/leipzig.html.

27. Fred Guteri, "The Truth About Global Warming: The Forecasts of Doom Are Mostly Guesswork, Richard Lindzen Argues—and He Has Bush's Ear," *Newsweek* (7/23/2001).

28. E. F. Mallove, "Lindzen Critical of Global Warming Prediction," *M.I.T. Tech Talk* (1989).

29. Richard S. Lindzen, "Global Warming: How to Approach the Science," Fourth International Conference on Climate Change (5/16/2010): *see* http://wattsupwiththat .files.wordpress.com/2010/05/lindzen_heartland_2010.pdf.

30. Richard S. Lindzen, "Climate of Fear: Global-warming Alarmists Intimidate Dissenting Scientists in Silence," GlobalResearch.ca (4/07/2007): *see* www.global research.ca/index.php?context=va&aid=5294/.

31. Tom Sharpe, "Former Astronaut Scoffs at Global Warming," *Santa Fe New Mexican* (2/14/2009): *see* www.santafenewmexican.com/SantaFeNorthernNM/Former-astronaut-scoffs-at-global-warming.

32. National Science Foundation: *see* www.nsf.gov/awardsearch/.

33. Richard S. Lindzen, Ming-Dah Chou, and Arthur Y. Hou, "Does the Earth Have an Adaptive Infrared Iris?" *Bulletin of the American Meteorological Society* 82 (2001): 417–32.

34. Richard S. Lindzen and Y. S. Choi, "On the Determination of Climate Feedbacks from Erbe Data," *Geophysical Research Letters* 36 (2009): L16705.

35. Kevin E. Trenberth et al, "Relationships between Tropical Sea Surface Temperature and Top-of-Atmosphere Radiation," *Geophysical Research Letters* 37 (2010): L03702.

36. Andrew C. Revkin, "A Rebuttal to a Cool Climate Paper," *New York Times* (1/08/2010): *see* http://dotearth.blogs.nytimes.com/2010/01/08/a-rebuttal-to-a-cool-climate-paper/.

37. "Former NASA Advisor Council Chair Jack Schmitt Quits Planetary Society Over New Roadmap," Space Ref (11/17/2008): *see* www.spaceref.com/news/viewsr .html?pid=29813.

38. Sharpe, "Former Astronaut Scoffs at Global Warming.

39. John Schwartz, "Vocal Minority Insists It Was All Smoke and Mirrors," *New York Times* (7/14/2009).

40. Nicholas Dawidoff, "The Global-Warming Heretic," *New York Times Magazine* (3/29/2009).

41. Freeman Dyson, "Heretical Thoughts About Science and Society," *Edge* (8/08/1987): *see* www.edge.org/3rd_culture/dysonf07/dysonf07_index.html.

42. Ann K. Finkbeiner, *The Jasons: The Secret History of Science's Postwar Elite*, 134. Subsequent references to this work are cited in the text.

43. Dyson, "Heretical Thoughts About Science and Society."

44. Ibid.

45. "Getting Heated," On the Media: NPR (4/10/2009): *see* www.onthemedia.org/ transcripts/2009/04/10/03.

46. "In a stunning journalistic lapse, the NY Times gives credulous coverage to Swift Boat smearer Marc Morano, the Jayson Blair of global warming," *Climate Progress* (4/09/2009): *see* http://climateprogress.org/2009/04/09/new-york-times-swift-boat-smearer-marc-morano-global-warming-denie-climatedepot/.

47. *See* http://bravethinkers.theatlantic.com/.

48. *See* www.sourcewatch.org/.

49. *See* www.tcsdaily.com/.

50. *See* www.desmogblog.com/.

51. *See* www.exxonsecrets.org/.

8. Fear of State: The Nonscientists

1. *See* http://mediamatters.org.

2. George Will, "Dark Green Doomsayers," *Washington Post* (2/15/2009).

3. J. D. Hays, John Imbrie, and N. J. Schackleton, "Variations in the Earth's Orbit: Pacemaker of the Ice Ages." *Science* 194 (1976): 1121–32.

4. John P. Holdren and Paul R. Ehrlich, *Global Ecology: Readings Toward a Rational Strategy for Man.*

5. Richard S. Lindzen, "Global Warming: The Origin and Nature of the Alleged Scientific Consensus," *Regulation*—The Cato Institute (1992): *see* www.cato.org/pubs/regulation/regv15n2/regi5n2g.html.

6. Thomas C. Peterson, William M. Connolley, and John Fleck, "The Myth of the 1970s Global Cooling Scientific Consensus," *Bulletin American Meteorological Association* 89 (2008): 1325–37.

7. "In reported response to Will controversy, *Wash. Post* ombudsman compounds global warming misinformation," *Media Matters* (2/22/2009): *see* http://mediamatters .org/research/200902220008.

8. El Niño comes from the Spanish and means "the little boy," or Christ Child, and refers to a periodic warming of the Eastern Pacific that sometimes occurs at Christmas. The opposite cooling effect is called La Niña; together the two make up the El Niño Southern Oscillation, which has a major effect on world weather.

9. George Will, "Climate Science in a Tornado," *Washington Post* (2/27/2009).

10. Andrew Alexander, "The Heat from a Global Warming Column," *Washington Post* (3/1/2009).

11. Chris Mooney, "Climate Change Myths and Facts," *Washington Post* (3/21/2009).

12. Michael Jarraud, "Understanding Climate Change," *Washington Post* (3/21/2009).

13. George Will, "Climate Change's Dim Bulbs," *Washington Post* (4/2/2009).

14. Juliet Eilperin, and Mary Beth Sheridan, "New Data Show Rapid Arctic Ice Decline: Proportion of Thicker, More Persistent Winter Cover Is the Lowest on Record," *Washington Post,* (4/7/2009).

15. Remarks to the Commonwealth Club by Michael Crichton, San Francisco, September 15, 2003; *see* www.monsanto.co.uk/news/ukshowlib.phtml?uid=7662.

16. Ronald Bailey, "A Chilling Tale: Michael Crichton's 'State of Fear.'" *Wall Street Journal* (12/10/2004).

17. Bruce Barcott, "'State of Fear': Not So Hot," *New York Times* (1/30/2005).

18. "Answers to Key Questions Raised by M. Crichton in 'State of Fear,'" Pew Center on Global Climate Change (8/17/2006): *see* www.pewclimate.org/state_of _fear.cfm.

201 | Notes

19. James Hansen, "Michael Crichton's 'Scientific Method'" (undated).

20. Seth Borenstein, "Novel on Global Warming Gets Some Scientists Burned Up," *Seattle Times* (2/20/2005).

21. Ibid.

22. "Michael Crichton's State of Confusion," *RealClimate* (12/13/2004): see www .realclimate.org/index.php/archives/2004/12/michael-crichtons-state-of-confusion/.

23. Michael Crichton, *State of Fear: A Novel.* (New York: HarperCollins: Avon Books, 2004), 636.

24. See www.globalwarmingisreal.com/.

25. Bjørn Lomborg, *The Skeptical Environmentalist: Measuring the Real State of the World.*

26. "The 'Skeptical Environmentalist.'" *The Economist* (9/6/2001).

27. David Pearce, "The Skeptical Environmentalist," *New Scientist* 171.2309 (9/22/2001): 50.

28. Denis Dutton, "Greener Than You Think," *Washington Post* (10/21/2001).

29. John Rennie, "Misleading Math About the Earth," *Scientific American* (2002).

30. Stuart Pimm and Jeff Harvey, "No Need to Worry About the Future," *Nature* 414 (2001): 149–50.

31. "UCS Examines 'The Skeptical Environmentalist,'" Union of Concerned Scientists: *see* www.ucsusa.org/global_warming/science_and_impacts/global_warming _contrarians/ucs-examines-the-skeptical.html.

32. Alex Kirby, "Bjorn Lomborg's Wonderful World," *BBC News* (8/23/2001): *see* http://news.bbc.co.uk/2/hi/uk_news/1502076.stm.

33. *See* www.lomborg-errors.dk.

34. Bjørn Lomborg, *Cool It: The Skeptical Environmentalist's Guide to Global Warming*, 8.

35. This section draws on a post by Michael Pawlyn on www.climateprogress.org.

36. Bjorn Lomborg, "Climate Change Decisions Should Be Based on Science, Not Political Activism," *The Guardian* (3/09/2009): *see* www.guardian.co.uk/environment/ cif-green/2009/mar/09/lomborg-climate-change.

37. Stefan Rahmstorf, "Climate Skeptics Confuse the Public by Focusing on Short-Term Fluctuations," *The Guardian* (3/9/2009).

38. N. H. Stern, *Stern Review on the Economics of Climate Change.*

39. "Profile: Bjorn Lomborg," *The Sunday Times* (8/9/2009); *see* www.timesonline .co.uk/tol/news/science/article6788630.ece.

40. Ibid.

41. Howard Friel, *The Lomborg Deception: Setting the Record Straight About Global Warming.*

42. Sharon Begley, "The Lomborg Deception: Debunking the Claims of the Climate-Change Skeptic," *Newsweek* (2/22/2010).

43. *See* http://coolit-themovie.com.

44. *See* www.heartland.org/events/NewYork09/proceedings.html.

45. Available at http://scienceandpublicpolicy.org/.

46. Tim Adams, "Monckton Saves the Day!" *The Observer* (5/6/2007).

47. George Monbiot, "This Is a Dazzling Debunking of Climate Change Science: It Is Also Wildly Wrong," *The Guardian* (11/14/2006).

48. For an image of the DVD cover, *see* http://scienceandpublicpolicy.org/apocalypseno-dvd.html.

49. Christopher Monckton, "The Myth of Heterosexual Aids," *The American Spectator*, (1/1987).

50. "TVMOB hate speech shocker: Lord Monckton repeats and expands on his charge that those who embrace climate science are 'Hitler youth' and fascists," *Climate Progress* (12/12/2009): *see* http://climateprogress.org/2009/12/12/tvmob-hate-speech-lord-monckton-hitler-youth-fascist-climate-activists/.

51. Lord Monckton of Brenchley, "Uphold Free Speech About Climate Change or Resign," Center for Science and Public Policy (12/11/2006): *see* http://ff.org/centers/csspp/pdf/20061212_monckton.pdf.

52. Judy Fahys, "Debate on Climate Heats up Online," *Salt Lake Tribune* (4/9/2010).

53. Monbiot, "This Is a Dazzling Debunking of Climate Change Science."

54. Available at http://scienceandpublicpolicy.org/.

55. *See* ibid.

56. "TVMOB hate speech shocker," *Climate Progress* (12/12/2009).

57. Adam Morton, "Climate Sceptic Clouds the Weather Issue," *The Age* (2/02/2010): *see* www.theage.com.au/environment/climate-change/climate-sceptic-clouds-the-weather-issue-20100201-n8y3.html.

58. *See* www.stthomas.edu/engineering/jpabraham/.

59. George Monboit, "Viscount Monckton, Another Fallen Idol of Climate Denial," *The Guardian* (6/03/2010): *see* www.guardian.co.uk/environment/georgemonbiot/2010/jun/03/monckton-climate-change.

60. John Abraham, "Abraham Reply to Monckton," *Skeptical Science* (6/6/2010); *see* www.skepticalscience.com/Abraham-reply-to-Monckton.html.

61. Christopher Monckton, "Climate: The Extremists Join the Debate at Last!" CFACT (6/4/2010);*see* http://cfact.eu/2010/06/04/climate-the-extremists-join-the-debate-at-last/.

62. Adams, "Monckton Saves the Day!"

63. Russell Baker, "Living Breathing Money," *New York Times* (12/6/1986).

9. Toxic Tanks

1. *See* www.fair.org/.

2. Union of Concerned Scientists, *Smoke, Mirrors & Hot Air*.

3. *See* http://www.sourcewatch.org.

4. Both quotations are from from www.sourcewatch.org.

5. *See* www.pewclimate.org/business/belc.

6. *See* http://www.sourcewatch.org.

7. "Getting Heated," *see* www.onthemedia.org/transcripts/2009/04/10/03.

8. Sometime in the fall of 2009, Heartland removed Bast's photo from its list of distinguished thinkers.

9. *See* www.sourcewatch.org.

10. Maureen Martin and Joseph L. Bast, "Welcome to Heartland's Smoker's Lounge!" The Heartland Institute (4/20/2007); see www.heartland.org/policybot/results/10594/Welcome_to_Heartlands_Smokers_Lounge.html.

11. Joseph L. Bast, "Five Lies about Tobacco: The Tobacco Bill Wasn't about Kids," The Heartland Institute (7/1/1998); see www.heartland.org/Article/732/July_1998_Five_Lies_about_Tobacco_The_Tobacco_Bill_Wasnt_about_Kids.html.

12. See www.heartland.org/about/truthsquad.html.

13. "Skeptic's Handbook spreads en masse: 150,000 copies!" JoNova; see http://joannenova.com.au/2009/03/skeptics-handbook-spreads-en-masse-150000-copies/.

14. B. D. Santer et al., "Consistency of Modelled and Observed Temperature Trends in the Tropical Troposphere," *International Journal of Climatology* 28 (2008): 1703–1722.

15. "Skeptic's Handbook spreads en masse: 150,000 copies!" JoNova.

16. Union of Concerned Scientists, *Smoke, Mirrors & Hot Air*, 36.

17. See www.exxonsecrets.org/.

18. W. Soon and S. Baliunas, "Proxy Climatic and Environmental Changes of the Past 1000 Years," *Climate Research* 23 (2003): 89–110.

19. "Revealed: Exxon Secret Funding of Global Warming Junk Scientists," Greenpeace (5/26/2009); see http://members.greenpeace.org/blog/exxonsecrets/2009/05/26/exxon_admits_2008_funding_of_global_warm.

20. "It's good work if you can get it," Rabett Run (5/8/2007); see http://rabett.blogspot.com/2007/05/its-good-work-if-you-can-get-it-in-1997.html.

21. See www.sourcewatch.org.

22. Ibid.

23. McCain, quoted in Chris Mooney, "Some Like It Hot," *Mother Jones* (May-June 2005).

24. See http://cei.org/about.

25. See www.sourcewatch.org.

26. Justin Bank, "Scientist to CEI: You Used My Research to Confuse and Mislead" (2006).

27. Chris Mooney, "An Inconvenient Assessment—Seven Years Ago, U.S. Scientists Published a Pioneering Study Detailing the Local Implications of Climate Change. Here's Why You've Never Heard of It." *Bulletin of the Atomic Scientists* 63 (2007): 40–47.

28. Rick Piltz, *Declaration in Support of Memorandum of Amici Curiae John F. Kerry and Jay Inslee* (2007).

29. President George W. Bush, quoted in Mooney, "An Inconvenient Assessment," 43.

30. "Myron Ebell" http://www.sourcewatch.org/index.php?title=Myron_Ebell.

31. Mooney, "An Inconvenient Assessment," 44.

32. Chris Mooney, *The Republican War on Science*.

33. See www.climatesciencewatch.org/.

34. M. G. Morgan et al., "Learning from the U.S. National Assessment of Climate Change Impacts," *Environmental Science & Technology* 39 (2005): 10.

35. Mooney, "An Inconvenient Assessment," 46.
36. *See* www.globalchange.gov/.
37. Michael Shayerson, "A Convenient Untruth," *Vanity Fair* (May 2007): 22.

10. An Industry to Trust

1. Union of Concerned Scientists, *Smoke, Mirrors & Hot Air*.
2. Ian Sample, "Scientists Offered Cash to Dispute Climate Study," *The Guardian* (2/2/2007).
3. "FACTSHEET: Citizens for a Sound Economy and CSE Educational Foundation, CSE"; *see* www.exxonsecrets.org/html/orgfactsheet.php?id=27.
4. In 2005, ExxonMobil made a profit of $36 billion.
5. Union of Concerned Scientists, *Smoke, Mirrors & Hot Air*.
6. Ibid., 40.
7. Ibid., 41.
8. Ibid., 20.
9. Ibid., 51.
10. Andrew Revkin, "Bush Aide Softened Greenhouse Gas Links to Global Warming," *New York Times* (6/8/2005).
11. *See* http://snowe.senate.gov/.
12. *See* www.climatesciencewatch.org/.
13. Ibid.
14. ExxonMobil, "2007 Corporate Citizenship Report"; *see* www.exxonmobil.com/Corporate/files/Corporate/community_ccr_2007.pdf.
15. David Adam, "ExxonMobil Continuing to Fund Climate Skeptic Groups, Records Show," *The Guardian* (7/1/2009).
16. *See* www.exxonmobil.com/.
17. Ario, quoted in Evan Lehmann, "Insurance Group Says Stolen E-Mails Show Risk in Accepting Climate Science" *New York Times* (1/13/2010).
18. Dominick V. Spracklen et al., "Atmospheric Science: Wildfires Drive Interannual Variability of Organic Carbon Aerosol in the Western U.S. in Summer," *Geophysical Research Letters* 34 (2009): D20301.
19. Buffet, quoted in Robert G. Hagstrom, *The Warren Buffett Portfolio: Mastering the Power of the Focus Investment Strategy*, 131.
20. L. Zuill and Ed Leefeldt, "U.S. Insurers Seen as Lagging on Global Warming," *Reuters* (3/9/2007).
21. Ibid.
22. Ibid.
23. Munich Re, *Highs and Lows: Weather Risks in Central Europe* (2008).
24. Swiss Re, *Opportunities and Risks of Climate Change* (2002).
25. Allianz, *Climate Change and Insurance* (2006).
26. Association of British Insurers, *Financial Risks of Climate Change* (2005).
27. National Association of Insurance Commissioners, "Climate Change and Global Warming (EX) Task Force"; *see* www.naic.org/committees_ex_climate.htm.

11. Balance as Bias: How the Media Missed
"The Story of the Century"

1. "The Story of the Century" used in the subtitle comes from Joseph J. Romm, *Hell and High Water: The Global Warming Solution*.

2. Boykoff, and Boykoff, "Balance as Bias."

3. Ibid., 134.

4. Sarah Palin, "The 'Cap and Tax' Dead End," *Washington Post* (7/14/2009).

5. Michael Dobbs, "Palin for Energy Czar!" The Fact Checker, *The Washington Post*; see http://voices.washingtonpost.com/fact-checker/2008/09/palin_on_energy .html.

6. Andrew Revkin, "In Climate Debate, Exaggeration Is a Pitfall," *New York Times* (2/23/2009).

7. Charles M. Blow, "Farewell, Fair Weather," *New York Times* (5/31/2008).

8. Quirin Schiermeier, "Insurers' Disaster Files Suggest Climate Is Culprit," *Nature* 441: 674.

9. Revkin, "In Climate Debate, Exaggeration Is a Pitfall."

10. Paul A. Offit, *Autism's False Prophets: Bad Science, Risky Medicine, and the Search for a Cure*, 177.

11. Ibid., 178.

12. Kalichman. *Denying Aids*.

13. Celia Farber, "Out of Control," *Harper's Magazine* 312 (March 2006).

14. Gail Beckerman, "*Harper's* Races Right Over the Edge of a Cliff," *Columbia Journalism Review* (3/8/2006); see www.cjr.org/behind_the_news/harpers_races_right_ over_the_e.php?page=1.

15. Alan Carlin and John Davidson, Proposed NCEE Comments on Draft Technical Support Document for Endangerment Analysis for Greenhouse Gas Emissions under the Clean Air Act. Draft, unpublished paper, 2009.

16. "Inhofe, Barrasso Question EPA Commitment to Transparency," U.S. Senate Committee on Environment and Public Works (7/1/2009); see http:// epw.senate.gov/public/index.cfm?FuseAction=Minority.PressReleases&Content Record_id=37d0fb4b-802a-23ad-440e-0fa4ee1a6506.

17. "Press Release: Barton: Global Warming Bill Will Send U.S. Back to 1875," House Energy and Commerce Committee Republicans (6/26/2009); see http:// republicans.energycommerce.house.gov/News/PRArticle.aspx?NewsID=7159.

18. Declan McCullagh, "EPA May Have Suppressed Report Skeptical of Global Warming" (6/26/2009); seewww.cbsnews.com/blogs/2009/06/26/politics/political hot sheet/entry5117890.shtml.

19. "Why the EPA Should Find Against 'Endangerment,'" World Climate Report (11/19/2008); see www.worldclimatereport.com/index.php/2008/11/19/why-the- epa-should-find-against-endangerment/.

20. McCullagh, "EPA May Have Suppressed Report Skeptical of Global Warming."

21. Society of Professional Journalists; see www.spj.org/ethicscode.asp.

22. Eric Pooley, *How Much Would You Pay to Save the Planet? The American Press and the Economics of Climate Change*; see www.hks.harvard.edu/presspol/publications/papers/discussion_papers/d49_pooley.pdf.

23. Eilperin and Sheridan, "New Data Show Rapid Arctic Ice Decline," *Washington Post* (4/7/2009).

24. "NY Times, WSJ, and Washington Post all rejected op-ed/letter from 255 National Academy of Sciences members defending climate science integrity," Climate Progress (5/13/2010); see http://climateprogress.org/2010/05/13/scientists-letter-on-climate-science-integrity-rejected-new-york-times-wsj-and-washington-post-polar-bear/.

12. Science Under Attack

1. See www.skepticalscience.com.

2. This discussion comes from Gerald A. Meehl et al., "Combinations of Natural and Anthropogenic Forcings in Twentieth-Century Climate," *Journal of Climate* 17 (2004): 3721–27.

3. Kevin E. Trenberth, Philip D. Jones, and Peter Ambenje, *IPCC Fourth Assessment Report—Chapter 3—Observations: Surface and Atmospheric Climate Change"* (2007), 244.

4. See www.sourcewatch.org.

5. National Oceanographic and Atmospheric Administration, *Talking Points Related to Concerns About Whether the U.S. Temperature Record Is Reliable* (2009).

6. Stephen McIntyre and Ross McKitrick, "Hockey Sticks, Principal Components, and Spurious Significance; Reply," *Geophysical Research Letters* 32 (2005): L03710.

7. S. McIntyre and R. McKitrick, "The M&M Critique of the MBH 98 Northern Hemispheric Climate Index: Update and Implications," *Energy & environment* 16 (2005): 69–100.

8. Joe Barton and Ed Whitfield, Letter to Dr. Raymond Bradley (6/23/2005); see http://republicans.energycommerce.house.gov/108/Letters/062305_Bradley.pdf.

9. Edward Wegman, David W. Scott, and Yasmin H. Said, *Ad Hoc Committee Report on The "Hockey Stick" Global Climate Reconstruction* (2006).

10. United States Congress, House Committee on Energy and Commerce, Subcommittee on Oversight and Investigations, *Questions Surrounding the "Hockey Stick" Temperature Studies: Implications for Climate Change Assessments: Hearings before the Subcommittee on Oversight and Investigations of the Committee on Energy and Commerce, House of Representatives.* 109th Cong., 2d sess., July 19 and July 27, 2006.

11. National Research Council, *Surface Temperature Reconstructions for the Last 2,000 Years.*

12. Richard Monastersky, "A Scientific Graph Stands Trial," *Chronicle of Higher Education* (9/6/2006).

13. E. R. Wahl and Caspar M. Ammann, "Robustness of the Mann, Bradley, Hughes Reconstruction," *Climatic Change* 85 (2007): 33–69.

14. Michael E. Mann et al., "Proxy-Based Reconstructions of Hemispheric and Global Surface Temperature Variations Over the Past Two Millennia," *Proceedings of the National Academy of Sciences of the United States of America* 105 (2008): 6.

15. Michael E. Mann et al., "Global Signatures and Dynamical Origins of the Little Ice Age and Medieval Climate Anomaly," *Science* 326 (2009): 1256–60.

16. James Hansen et al., "Potential Climate Impact of Mount Pinatubo Eruption," *Geophysical Research Letters* 19 (1992): 215–18.

17. D. S. Kaufman et al., "Recent Warming Reverses Long-Term Arctic Cooling," *Science* 325 (2009): 1236–39.

18. Broecker and Kunzig, *Fixing Climate*, 185.

19. "Is Antarctica Losing or Gaining Ice?" Skeptical Science; *see* www.skeptical science.com/antarctica-gaining-ice.htm.

20. "Satellite Temperature Measurements" Wikipedia; *see* http://en.wikipedia.org/wiki/Satellite_temperature_measurements.

13. Greatest Hoax in History?

1. James Inhofe, "Climate Change Update, Senate Floor Statement by U.S. Sen. James M. Inhofe (R-Okla)" (1/4/2005); *see* http://inhofe.senate.gov/pressreleases/climateupdate.htm.

2. James Inhofe, "Global Warming 'Consensus' in Freefall: More Than 650 International Scientists Dissent over Man-Made Global Warming Claims." Speech delivered on Senate Floor (1/8/2009).

3. Louise Roug, "Jim Inhofe Gets Cool Reception in Denmark," *Politico* (12/19/2009); *see* www.politico.com/news/stories/1209/30769.html.

4. Jesse O. Kurtz, "Stephen Moore Addresses New Jersey Citizen Activists," Conservative Majority New Jersey (5/31/2009); *see* www.gopusanj.com/wordpress/?p=6632.

5. Harold Ambler, "Mr. Gore: Apology Accepted," *Huffington Post* (1/3/2009); *see* www.huffingtonpost.com/harold-ambler/mr-gore-apology-accepted_b_154982 .html.

6. Lutz, quoted in Lascelles Linton, "GM CEO Responds to Lutz's Global Warming's a 'total crock of sh*t' comment," Autobloggreen (3/11/2008); *see* www.autoblog green.com/2008/03/11/gm-ceo-responds-to-lutzs-global-warmings-a-total-crock-of-sh/.

7. John Coleman and Joe D'Aleo, "Weather Channel Founder: Global Warming 'Greatest Scam in History,'" It's Rainmaking Time (11/11/2007); *see* http://icecap.us/index.php/go/joes-blog/comments_about_global_warming/.

8. *See* www.exxonsecrets.org/.

9. Harrison H. Schmitt, "We'd Better Be Right On Climate Science," The Heartland Institute (5/7/2009); *see* www.heartland.org/full/25282/Wed_Better_Be_Right_On_Climate_Science.html.

10. *The Economist*, "Monkey Business? Allegations of Scientific Misconduct at Harvard Have Academics up in Arms" (8/26/2010); *see* www.economist.com/node/16886218.

11. Ibid.

12. Nicholas Wade, "Harvard Researcher May Have Fabricated Data," *New York Times* (8/27/2010).

13. Jeffrey Mervis, "An Inside/Outside View of U.S. Science," *Science* 325 (July 10, 2009): 132–33.

14. Richard Lindzen, "The Press Gets It Wrong: Our Report Doesn't Support the Kyoto Treaty," *Wall Street Journal* (6/11/2001).

15. *See* www.heartland.org/events/NewYork09/proceedings.html.

16. Union of Concerned Scientists; *see* www.ucsusa.org/scientific_integrity/.

17. Bergmann, "Rep. Broun Receives Applause on the House Floor for Calling Global Warming a 'Hoax,'"; *see* http://thinkprogress.org/2009/06/26/broun-globalwarming-hoax.

18. Thomas Fingar, *National Intelligence Assessment on the National Security Implications of Global Climate Change to 2030* (2008).

19. Military Advisory Board, *National Security and the Threat of Climate Change* (2007).

20. *See* www.defense.gov/QDR/.

14. Climategate: Much Ado About Nothing

1. *The Guardian*, "Copenhagen Climate Change Conference: 'Fourteen days to seal history's judgment on this generation'" (12/7/2009); *see* www.guardian.co.uk/commentisfree/2009/dec/06/copenhagen-editorial.

2. *See* www.eastangliaemails.com/index.php.

3. James Delingpole, "Climategate: The Final Nail in the Coffin of 'Anthropogenic Global Warming'?" *Telegraph* (11/20/2009); *see* http://blogs.telegraph.co.uk/news/jamesdelingpole/100017393/climategate-the-final-nail-in-the-coffin-of-anthropogenic-global-warming/.

4. Patrick Michaels, "How to Manufacture a Climate Consensus," *Wall Street Journal* (12/17/2009); *see* http://online.wsj.com/article/SB10001424052748704398304574598230426037244.html.

5. Alex Seitz-Wald, "Inhofe's Hoax: Senator Distorts Meteorological Study to Show Support for His Global Warming Denial," *Think Progress* (12/8/2009); *see* http://thinkprogress.org/2009/12/08/inhofe-mischaracterizes-meteorologial/.

6. Pachauri, quoted in "Climatic Research Unit Email Controversy," Wikipedia; *see* http://en.wikipedia.org/wiki/Climategate#cite_ref-UEA_24_Nov_28-0.

7. Delingpole, "Climategate."

8. Kevin E. Trenberth, "An Imperative for Climate Change Planning: Tracking Earth's Global Energy," *Current Opinion in Environmental Sustainability* 1 (2009): 19–27.

9. "Climatic Research Unit Email Controversy, AGU Statement Regarding the Recent Release of E-mails Hacked from the Climate Research Unit at University of East Anglia," American Geophysical Union; *see* www.agu.org/news/archives/2009–12–08_hacked-emails-climate-researchshtml.shtml.

10. Leshner, quoted in "AAAS Reaffirms Statements on Climate Change and Integrity," Advancing Science, Serving Society; see www.aaas.org/news/releases/2009/1204climate_statement.shtml.

11. "Climatologists Under Pressure," *Nature* 462.25 (January 2010).

12. "Contrarians Using Hacked E-mails to Attack Climate Science," Union of Concerned Scientists (11/23/2009); see www.ucsusa.org/news/press_release/hacked-climate-e-mails-0306.html.

13. Henry Foley et al., "RA-10 Inquiry Report: Concerning the Allegations of Research Misconduct Against Dr. Michael E. Mann, Department of Meteorology, College of Earth and Mineral Sciences, The Pennsylvania State University" (2/3/2010); see www.research.psu.edu/orp/Findings_Mann_Inquiry.pdf.

14. Sarah Assmann et al., "RA-10 Final Investigation Report Involving Dr. Michael E. Mann, The Pennsylvania State University" (6/4/2010); see http://live.psu.edu/fullimg/userpics/10026/Final_Investigation_Report.pdf.

15. House of Commons Science and Technology Committee, "The Disclosure of Climate Data from the Climatic Research Unit at the University of East Anglia" (3/24/2010); see www.publications.parliament.uk/pa/cm200910/cmselect/cmsctech/387/387i.pdf.

16. "Report of the International Panel set up by the University of East Anglia to examine the research of the Climatic Research Unit"; see www.uea.ac.uk/mac/comm/media/press/CRUstatements/SAP.

17. Sir Muir Russell et al., "The independent Climate Change E-mails Review," July 2010; see www.cce-review.org/pdf/FINAL%20REPORT.pdf.

18. See www.official-documents.gov.uk/documents/cm79/7934/7934.pdf.

19. IPCC, "10.6.2 The Himalayan Glaciers," *IPCC Fourth Assessment Report: Climate Change 2007* (New York: UN Intergovermental Panel on Climate Change, 2007); see www.ipcc.ch/publications_and_data/ar4/wg2/en/ch10s10-6-2.html.

20. For a discussion of the provenance of "Glaciergate," see Bidisha Banerjee and George Collins, "Anatomy of IPCC's Mistake on Himalayan Glaciers and Year 2035," The Yale Forum on Climate Change and The Media (2/4/2010); see www.yaleclimatemediaforum.org/2010/02/anatomy-of-ipccs-himalayan-glacier-year-2035-mess/.

21. Graciela Magrin et al., "Latin America," *IPCC Fourth Assessment Report: Climate Change 2007* (New York: UN Intergovermental Panel on Climate Change, 2007); see www.ipcc.ch/pdf/assessment-report/ar4/wg2/ar4-wg2-chapter13.pdf, 596.

22. James Delingpole, "After Climategate, Pachaurigate and Glaciergate: Amazongate" *The Telegraph* (1/25/2010); see http://blogs.telegraph.co.uk/news/jamesdelingpole/100023598/after-climategate-pachaurigate-and-glaciergate-amazongate/.

23. George Monbiot, "Sunday Times Admits 'Amazongate' Story Was Rubbish. But Who's to Blame?" *The Guardian* (6/24/2010); see www.guardian.co.uk/environment/georgemonbiot/2010/jun/24/sunday-times-amazongate-ipcc.

24. InterAcademy Council Review of the IPCC; see http://reviewipcc.interacademycouncil.net/.

25. Jeremy A. Kaplan, "Independent Audit Panel Slams U.N.'s Climate Group, Fox News (8/30/2010); see www.foxnews.com/scitech/2010/08/30/independent-audit-slams-un-climate-panel/?test=latestnews.

15. Anatomy of Denial

1. Du Pont set up, then knocked down, his global warming straw man: "[I]t is not clear that human activity is wholly responsible," MediaMatters (3/28/2006); *see* http://mediamatters.org/research/200603280010.

2. Shayerson, "A Convenient Untruth," *Vanity Fair* (May 2007).

3. The Luntz memo is available at www.ewg.org/node/8684.

4. "Frank Luntz Explains 'Words That Work,'" Interview with Terry Gross, *Fresh Air* (1/9/2007); *see* www.npr.org/templates/story/story.php?storyId=6761960.

5. George Orwell, *Nineteen Eighty-Four: A Novel*.

6. "Lysenkoism," Wikipedia; *see* http://en.wikipedia.org/wiki/Lysenkoism. See also Loren Graham, *What Have We Learned About Science and Technology from the Russian Experience?* (Stanford: Stanford UP, 1998), 19.

7. "Glenn Talks with Lord Monckton," *The Glenn Beck Program* (3/2/2008); *see* www.glennbeck.com/content/articles/article/196/6783/.

8. David Joravsky, *The Lysenko Affair*, 58.

9. Christina Larson and Joshua Keating, "The FP Guide to Climate Skeptics, *Foreign Policy*, February 26, 2010; *see* http://www.foreignpolicy.com/articles/2010/02/25/the_fp_guide_to_climate_skeptics

10. Debora MacKenzie, "Living in Denial: Why Sensible People Reject the Truth," *New Scientist* (5/19/2010).

16. Escalating Tactics

1. Steve Crabtree, "New Poll Gauges Americans' General Knowledge Levels," *Gallup* (7/6/1999); *see* www.gallup.com/poll/3742/new-poll-gauges-americans-general-knowledge-levels.aspx.

2. *See* www.intelligentdesign.org.

3. Associated Press, "Judge Rules Against 'Intelligent Design,'" MSNBC (12/20/2005); *see* http://www.msnbc.msn.com/id/10545387/. You can hear Judge Jones read from his decision at www.pbs.org/wgbh/nova/evolution/judge-speaks.html.

4. Louisiana Senate Bill No. 733, Act No. 473, Regular Session, 2008; *see* http://ncse.com/webfm_send/792.

5. *See* http://ncse.com/evolution/top-ten-evolution-stories-2010.

6. "House Recurrent Resolution No. 1009," State of South Dakota, 85th sess., Legislative Assembly, 2010; *see* http://legis.state.sd.us/sessions/2010/Bill.aspx?File=HCR1009P.htm.

7. "CID No. 2-MM, Civil Investigations Demand"; *see* http://voices.washington post.com/virginiapolitics/Virginia_Attorney_General_CID.pdf.

8. Ann B. Hamric and University of Virginia Faculty Senate Executive Council, "Position Statement on Attorney General's Investigation of Dr. Michael Mann," (5/5/2010); *see* http://voices.washingtonpost.com/virginiapolitics/UVa%20Faculty%20Senate.pdf.

9. "UVA Files Petition Against Cuccinelli Demands," NBC (5/27/2010); *see* www .nbc29.com/global/story.asp?s=12555610.

10. Commonwealth of Virgina, "The Rector and Visitors of the University of Virgina v. Kenneth T. Cuccinelli, II, Attorney General of Viginia, Case No.: CL10–398" (8/30/2010); *see* http://voices.washingtonpost.com/virginiapolitics/2010–08–30%20 Opinion%20Granting%20UVA%20Petition.pdf.

11. *See* http://voices.washingtonpost.com/virginiapolitics/New%20Mann%20 CID.pdf.

Bibliography

Adam, David. "ExxonMobil Continuing to Fund Climate Skeptic Groups, Records Show." *The Guardian*, July 1, 2009.

Adams, Tim. "Monckton Saves the Day!" *The Observer*, May 6, 2007.

Alexander, Andrew. "The Heat from a Global Warming Column." *Washington Post*, March 1, 2009.

Allianz. *Climate Change and Insurance*. Munich, 2006.

Anderegg, William R. L., James W. Prall, Jacob Harold, and Stephen H. Schneider. "Expert Credibility in Climate Change." *Proceedings of the National Academy of Sciences of the United States of America* 107.27 (2010): 12107–12109.

Arrhenius, Svante and H. Borns. *Worlds in the Making: The Evolution of the Universe*. New York and London: Harper, 1908.

Association of British Insurers. *Financial Risks of Climate Change* (2005; www.abi .org.uk/Publications/ABI_Publications_Financial_Risks_of_Climate_Change_do7 .aspx).

Bailey, Ronald. "A Chilling Tale: Michael Crichton's 'State of Fear.'" *Wall Street Journal*, December 10, 2004.

Baker, Russell. "Living Breathing Money." *New York Times*, December 6, 1986.

Bank, Justin. "Scientist to CEI: You Used My Research to Confuse and Mislead" (2006; www.sourcewatch.org/index.php?title=Competitive_Enterprise_Institute _And_Global_Warming).

Barcott, Bruce. "'State of Fear': Not So Hot." *New York Times*, January 30, 2005.

Begley, Sharon. "The Lomborg Deception: Debunking the Claims of the Climate-Change Skeptic." *Newsweek*, February 22, 2010.

Blow, Charles M. "Farewell, Fair Weather." *New York Times*, May 31, 2008.

Borenstein, Seth. "Novel on Global Warming Gets Some Scientists Burned Up." *Seattle Times*, February 20, 2005.

Boykoff, Maxwell T. and Jules M. Boykoff. "Balance as Bias: Global Warming and the US Prestige Press." *Global Environmental Change: Human and Policy Dimensions* 14 (2004): 125–36.

Broecker, Wallace S. "Climatic Change: Are We on the Brink of a Pronounced Global Warming?" *Science* 189 (1975): 460–63.

———. "Shifting Rainfall: A Paleo Perspective." *EOS Trans. AGU* 89.53 (Fall Mtg. Supp., Abstract UA22A-05, 2008).

Broecker, Wallace S. and Robert Kunzig. *Fixing Climate: What Past Climate Changes Reveal About the Current Threat—and How to Counter It.* New York: Hill and Wang, 2009.

Callendar, G. S. "The Artificial Production of Carbon Dioxide and Its Influence on Climate." *Quarterly Journal of the Royal Meteorological Association* 64 (1938): 223–40.

Carlin, Alan and John Davidson. Proposed NCEE Comments on Draft Technical Support Document for Endangerment Analysis for Greenhouse Gas Emissions under the Clean Air Act. Draft. Unpublished paper, 2009.

Crichton, Michael. *State of Fear: A Novel.* New York: HarperCollins: Avon Books, 2004.

Dawidoff, Nicholas. "The Global-Warming Heretic." *New York Times Magazine*, March 29, 2009.

Doran, Peter and Maggie Kendall Zimmerman. "Examining the Scientific Consensus on Climate Change." *EOS* 90 (2009): 22–23.

Dutton, Denis. "Greener Than You Think." *Washington Post*, October 21, 2001.

The Economist. "Monkey Business? Allegations of Scientific Misconduct at Harvard Have Academics up in Arms," August 26, 2010.

Eilperin, Juliet and Mary Beth Sheridan. "New Data Show Rapid Arctic Ice Decline: Proportion of Thicker, More Persistent Winter Cover Is the Lowest on Record." *Washington Post*, April 7, 2009.

Einstein, Albert. "Does the Inertia of a Body Depend Upon Its Energy Content?" *Annalen der Physik* 18 (1905): 639–41.

Fahys, Judy. "Debate on Climate Heats up Online." *Salt Lake Tribune*, April 9, 2010.

Farber, Celia. "Out of Control." *Harper's Magazine* 312 (March 2006).

Feynman, Richard. "Personal Observations on the Reliability of the Shuttle," *Appendix F, NASA Challenger Report* (1986); see http://science.ksc.nasa.gov/shuttle/ missions/51-l/docs/rogers-commission/Appendix-F.txt.

Fingar, Thomas. *National Intelligence Assessment on the National Security Implications of Global Climate Change to 2030* (2008; www.dni.gov/testimonies/20080625_ testimony.pdf).

Finkbeiner, Ann K. *The Jasons: The Secret History of Science's Postwar Elite.* New York: Penguin, 2007.

Friel, Howard. *The Lomborg Deception: Setting the Record Straight About Global Warming.* New Haven: Yale UP, 2010.

Gelbspan, Ross. *The Heat Is On: The Climate Crisis, the Cover-up, the Prescription.* Updated ed. Reading, Mass.: Perseus, 1998.

———. "The Heat Is On: The Warming of the World's Climate Sparks a Blaze of Denial." *Harper's* 291 (1995): 9.

Gleick, Peter and 254 others. "Climate Change and the Integrity of Science." *Science* 328 (2010): 689–90.

Guteri, Fred. "The Truth About Global Warming: The Forecasts of Doom Are Mostly Guesswork, Richard Lindzen Argues—and He Has Bush's Ear." *Newsweek*, July 23, 2001.

Hagstrom, Robert G. *The Warren Buffett Portfolio: Mastering the Power of the Focus Investment Strategy.* New York: Wiley, 1999.

Hansen, James. "Michael Crichton's 'Scientific Method'" (undated; www.columbia .edu/~jeh1_2005/Crichton_20050927.pdf).

Hansen, James et al. "Potential Climate Impact of Mount Pinatubo Eruption." *Geophysical Research Letters* 19 (1992): 215–18.

Hays, J. D., John Imbrie, and N. J. Schackleton. "Variations in the Earth's Orbit: Pacemaker of the Ice Ages." *Science* 194 (1976): 1121–32.

Hebert, H. J. "Jokers Add Fake Names to Warming Petition." *Seattle Times*, May 1, 1998.

Hertsgaard, Mark. "While Washington Slept." *Vanity Fair* (May 2006).

Holdren,John P. and Paul R. Ehrlich. *Global Ecology: Readings Toward a Rational Strategy for Man.* New York: Harcourt Brace Jovanovich, 1971.

Huxley, Thomas Henry. *Selected Essays and Addresses of Thomas Henry Huxley.* Edited by Philo Melvyn Buck, Jr. New York: Macmillan, 1910.

Inhofe, James. "Global Warming 'Consensus' in Freefall: More Than 650 International Scientists Dissent over Man-Made Global Warming Claims." Speech delivered on Senate Floor, January 8, 2009.

Jaffe, Bernard. *Crucibles: The Story of Chemistry from Ancient Alchemy to Nuclear Fission.* New rev. and updated 4th ed. New York: Dover, 1976.

Jarraud, Michael. "Understanding Climate Change." *Washington Post*, March 21, 2009.

Joravsky, David. *The Lysenko Affair.* Chicago: U of Chicago P, 1970.

Kalichman, Seth C. *Denying Aids: Conspiracy Theories, Pseudoscience, and Human Tragedy.* New York: Copernicus Books, 2009.

Kaufman, D. S. et al. "Recent Warming Reverses Long-Term Arctic Cooling." *Science* 325 (2009): 1236–39.

Kennedy, Donald. "Editorial—an Unfortunate U-Turn on Carbon." *Science* 291 (2001): 2.

Kerr, Richard A. "Clouds Appear to Be Big, Bad Player in Global Warming." *Science* 325 (2009): 376.

Lehmann, Evan. "Insurance Group Says Stolen E-Mails Show Risk in Accepting Climate Science." *New York Times*, January 13, 2010; www.nytimes.com/ cwire/2010/01/13/13climatewire-insurance-group-says-stolen-e-mails-show-ris-91554.html.

Lindzen, Richard. "The Press Gets It Wrong: Our Report Doesn't Support the Kyoto Treaty." *Wall Street Journal*, June 11, 2001.

Lindzen, Richard S., Ming-Dah Chou, and Arthur Y. Hou. "Does the Earth Have an Adaptive Infrared Iris?" *Bulletin of the American Meteorological Society* 82 (2001): 417–32.

Lindzen, Richard S. and Y. S. Choi. "On the Determination of Climate Feedbacks from Erbe Data." *Geophysical Research Letters* 36 (2009): L16705.

Lomborg, Bjørn. *Cool It: The Skeptical Environmentalist's Guide to Global Warming.* New York: Knopf, 2007.

———. *The Skeptical Environmentalist: Measuring the Real State of the World*. Cambridge and New York: Cambridge UP, 2001.

MacKenzie, Debora. "Living in Denial: Why Sensible People Reject the Truth." *New Scientist*, May 19, 2010.

Malakoff, David. "Climate Change: Advocacy Mailing Draws Fire." *Science* 280 (1998): 195.

Mallove, E. F. "Lindzen Critical of Global Warming Prediction." *M.I.T. Tech Talk* (1989).

Mann, M. E., R. S. Bradley, and M. K. Hughes. "Global-Scale Temperature Patterns and Climate Forcing over the Past Six Centuries." *Nature* 392 (1998): 779–87.

Mann, Michael E. et al. "Global Signatures and Dynamical Origins of the Little Ice Age and Medieval Climate Anomaly." *Science* 326 (2009): 1256–60.

———. "Proxy-Based Reconstructions of Hemispheric and Global Surface Temperature Variations Over the Past Two Millennia." *Proceedings of the National Academy of Sciences of the United States of America* 105 (2008): 6.

Matthews, R. "Leading Scientific Journals 'Are Censoring Debate on Global Warming.'" *The Telegraph*, May 1, 2005.

McIntyre, Stephen and Ross McKitrick. "Hockey Sticks, Principal Components, and Spurious Significance; Reply." *Geophysical Research Letters* 32 (2005): L03710.

———. "The M&M Critique of the MBH 98 Northern Hemispheric Climate Index: Update and Implications." *Energy & environment* 16 (2005): 69–100.

Meehl, Gerald A. et al. "Combinations of Natural and Anthropogenic Forcings in Twentieth-Century Climate." *Journal of Climate* 17 (2004): 3721–27.

Mervis, Jeffrey. "An Inside/Outside View of U.S. Science," *Science* 325 (July 10, 2009): 132–33.

Michaels, David. *Doubt Is Their Product: How Industry's Assault on Science Threatens Your Health*. New York: Oxford UP, 2008.

Military Advisory Board. *National Security and the Threat of Climate Change*. Alexandria, Va.: The CNA Corporation, 2007.

Monastersky, Richard. "A Scientific Graph Stands Trial." *Chronicle of Higher Education*, September 6, 2006.

Monbiot, George. "This Is a Dazzling Debunking of Climate Change Science: It Is Also Wildly Wrong." *The Guardian*, November 14, 2006.

Monckton, Christopher. "'Consensus'? What 'Consensus'? Among Climate Scientists, the Debate Is Not Over." Science and Public Policy Institute, 2007.

———. "The Myth of Heterosexual Aids." *The American Spectator*, January 1987.

Mooney, Chris. "Climate Change Myths and Facts." *Washington Post*, March 21, 2009.

———. "An Inconvenient Assessment—Seven Years Ago, U.S. Scientists Published a Pioneering Study Detailing the Local Implications of Climate Change. Here's Why You've Never Heard of It." *Bulletin of the Atomic Scientists* 63 (2007): 40–47.

———. *The Republican War on Science*. New York: Basic Books, 2005.

———. "Some Like It Hot." *Mother Jones* (May–June 2005).

Morgan, M. G. et al. "Learning from the U.S. National Assessment of Climate Change Impacts." *Environmental Science & Technology* 39 (2005): 10.

Munich Re. *Highs and Lows: Weather Risks in Central Europe*. Munich, 2008.

National Oceanographic and Atmospheric Administration. *Talking Points Related to Concerns About Whether the U.S. Temperature Record Is Reliable*. Washington, D.C.: NOAA, 2009.

National Research Council. *Advancing the Science of Climate Change*. Washington, D.C.: National Academy Press, 2010.

———. *Climate Change Science: An Analysis of Some Key Questions*. Washington, D.C.: National Academy Press, 2001.

———. *Surface Temperature Reconstructions for the Last 2,000 Years*. Washington, D.C.: National Academies Press, 2006.

National Research Council, Climate Research Board. *Carbon Dioxide and Climate: A Scientific Assessment*. Washington, D.C.: National Academy of Sciences, 1979.

National Research Council, Committee on Atmospheric Sciences. *Weather and Climate Modification Problems and Prospects: Final Report of the Panel on Weather and Climate Modification*. Washington, D.C.: National Academy of Sciences/National Research Council, 1966.

Nobel Laureates. "The St. James's Palace Memorandum: 'Action for a Low Carbon and Equitable Future.'" London, UK, May 26 – 28, 2009; www.nobel-cause.de/SJP_Memorandum_english.pdf.

Offit, Paul A. *Autism's False Prophets: Bad Science, Risky Medicine, and the Search for a Cure*. New York: Columbia UP, 2008.

Oreskes, Naomi. "The Long Consensus on Global Warming." *Washington Post*, February 19, 2007.

———. "Science and Public Policy: What's Proof Got to Do with It?" *Environmental Science & Policy* 7 (2004): 369–83.

———. "The Scientific Consensus on Climate Change." *Science* 306 (2004): 1686.

Oreskes, Naomi and E. M. Conway. *Merchants of Doubt: How a Handful of Scientists Obscured the Truth on Issues from Tobacco to Global Warming*. New York: Bloomsbury, 2010.

Orwell, George. *Nineteen Eighty-Four: A Novel* (1949). New York: Plume, 1983.

Oxford University Press. *The Oxford Dictionary of Quotations*. 3d ed. Oxford and New York: Oxford UP, 1979.

Palin, Sarah. "The 'Cap and Tax' Dead End." *Washington Post*, July 14, 2009.

Pearce, David. "The Skeptical Environmentalist." *New Scientist* 171.2309 (September 22, 2001): 50.

Peterson, Thomas C., William M. Connolley, and John Fleck. "The Myth of the 1970s Global Cooling Scientific Consensus." *Bulletin American Meteorological Association* 89 (2008): 1325–37.

Piltz, Rick. *Declaration in Support of Memorandum of Amici Curiae John F. Kerry and Jay Inslee, in Center for Biological Diversity et al. v. Dr. William Brennan* (February 7, 2007); www.law.stanford.edu/program/clinics/environmental/Piltz_Declaration_Final.pdf.

Pimm, Stuart and Jeff Harvey. "No Need to Worry About the Future." *Nature* 414 (2001): 149–50.

Pooley, Eric. *How Much Would You Pay to Save the Planet? The American Press and the Economics of Climate Change*. Cambridge: Harvard University Press and the John F. Kennedy School of Government, 2009.

President's Science Advisory Committee, Environmental Pollution Panel (White House, 1965). *Restoring the Quality of Our Environment.* Washington, D.C.: GPO, 1965.

"Profile: Bjorn Lomborg." *The Sunday Times*, August 9, 2009.

Rahmstorf, Stefan. "Climate Skeptics Confuse the Public by Focusing on Short-Term Fluctuations." *The Guardian*, March 9, 2009.

Rennie, John. "Misleading Math About the Earth." *Scientific American* (2002).

Revelle, Roger and Hans E. Suess. "Carbon Dioxide Exchange between Atmosphere and Ocean and the Question of an Increase of Atmospheric CO_2 During the Past Decades." *Tellus* 9 (1957): 18–27.

Revkin, Andrew. "Bush Aide Softened Greenhouse Gas Links to Global Warming." *New York Times*, June 8, 2005.

———. "In Climate Debate, Exaggeration Is a Pitfall." *New York Times*, February 23, 2009.

Romm, Joseph J. *Hell and High Water: The Global Warming Solution.* 1st Harper Perennial ed. New York: Harper Perennial, 2008.

Sample, Ian. "Scientists Offered Cash to Dispute Climate Study." *The Guardian*, February 2, 2007.

Santer, B. D., et al. "Consistency of Modelled and Observed Temperature Trends in the Tropical Troposphere." *International Journal of Climatology* 28 (2008): 1703–1722.

Schiermeier, Quirin. "Insurers' Disaster Files Suggest Climate Is Culprit." *Nature* 441: 674.

Schwartz, John. "Vocal Minority Insists It Was All Smoke and Mirrors." *New York Times*, July 14, 2009.

Shayerson, Michael. "A Convenient Untruth." *Vanity Fair* (May 2007): 22.

Showstack, R. "Science Academies' Statement on Climate Change." *EOS* 90 (2009): 216.

Singer, S. Fred. "Anthology of 1995's Environmental Myths," *Washington Times*, February 11, 1996.

"The 'Skeptical Environmentalist.'" *The Economist*, September 6, 2001.

Solomon, Susan. *Climate Change 2007: The Physical Science Basis; Summary for Policymakers, Technical Summary and Frequently Asked Questions. Part of the Working Group I Contribution to the Fourth Assessment Report of the Intergovernmental Panel on Climate Change.* 1st published ed. Nairobi: UNEP, 2007.

Soon, W. and S. Baliunas. "Proxy Climatic and Environmental Changes of the Past 1000 Years." *Climate Research* 23 (2003): 89–110.

Spracklen, Dominick V. et al. "Atmospheric Science: Wildfires Drive Interannual Variability of Organic Carbon Aerosol in the Western U.S. in Summer." *Geophysical Research Letters* 34 (2009): D20301.

Stern, N. H. *Stern Review on the Economics of Climate Change.* London: HM Treasury, 2006.

Swiss Re. *Opportunities and Risks of Climate Change.* Zurich, 2002.

Tierney, John. "The Big City: It's a Sin." *New York Times*, April 9, 1995.

———. "Flawed Science Advice for Obama?" *New York Times*, December 19, 2008.

Trenberth, Kevin E. "An Imperative for Climate Change Planning: Tracking Earth's Global Energy." *Current Opinion in Environmental Sustainability* 1 (2009): 19–27.

Trenberth, Kevin E. et al. "Relationships between Tropical Sea Surface Temperature and Top-of-Atmosphere Radiation." *Geophysical Research Letters* 37 (2010): L03702.

Trenberth, Kevin E., Philip D. Jones, and Peter Ambenje. *IPCC Fourth Assessment Report—Chapter 3—"Observations: Surface and Atmospheric Climate Change."* New York: UN Intergovernmental Panel on Climate Change, 2007.

Tyndall, John. "On the Absorption and Radiation of Heat by Gases and Vapours, and on the Physical Connexion of Radiation, Absorption, and Conduction." *Philosophical Magazine* 22 (1861): 169–94, 273–85.

Union of Concerned Scientists. *Smoke, Mirrors & Hot Air: How Exxonmobil Uses Big Tobacco's Tactics to Manufacture Uncertainty on Climate Science.* Cambridge, Mass.: Union of Concerned Scientists, 2007.

United States Congress. House Committee on Energy and Commerce. Subcommittee on Oversight and Investigations. Questions Surrounding the "Hockey Stick" Temperature Studies: Implications for Climate Change Assessments: Hearings before the Subcommittee on Oversight and Investigations of the Committee on Energy and Commerce, House of Representatives. 109th Cong., 2d sess., July 19 and July 27, 2006.

Wade, Nicholas. "Harvard Researcher May Have Fabricated Data." *New York Times*, August 27, 2010.

Wahl, E. R. and Caspar M. Ammann. "Robustness of the Mann, Bradley, Hughes Reconstruction." *Climatic Change* 85 (2007): 33–69.

Weart, Spencer R. *The Discovery of Global Warming.* Rev. and expanded ed. Cambridge: Harvard UP, 2008.

Wegman, Edward, David W. Scott, and Yasmin H. Said. *Ad Hoc Committee Report on The "Hockey Stick" Global Climate Reconstruction for the Chairmen of the U.S. House Committee on Energy and Commerce of the Subcommittee on Oversight and Investigations* (2006); *see* www.uoguelph.ca/~rmckitri/research/WegmanReport.pdf.

Will, George. "Al Gore's Green Guilt," *Washington Post*, September 3, 1992.

———. "Climate Change's Dim Bulbs." *Washington Post*, April 2, 2009.

———. "Climate Science in a Tornado." *Washington Post*, February 27, 2009.

———. "Dark Green Doomsayers." *Washington Post*, February 15, 2009.

Zuill, L. and Ed Leefeldt. "U.S. Insurers Seen as Lagging on Global Warming." *Reuters*, March 9, 2007.

Acknowledgments

Thanks to Robert Christopherson, Joan Hartmann, Jon Kussmaul, Joanna Powell, agent John Thornton, and four anonymous reviewers selected by Columbia University Press for reading the manuscript and making many improvements. Special thanks to Sean Pool, who diligently read the entire manuscript and made many useful comments. Patrick Fitzgerald, Publisher for Life Sciences at Columbia University Press, gave the manuscript a close reading and improved it significantly. The team at Columbia University Press was a delight to work with and made many substantial contributions. Thanks also to the intrepid scientists at ClimateProgress.org, RealClimate .org, and SkepticalScience.com, from whom I have learned so much. People of the future will laud you as heroes.

Index

Initial articles in titles (A, An, The) are ignored in sorting. Page numbers followed by f indicate figures, and *n* indicates note.